Synthesis Lectures on Learning, Networks, and Algorithms

Series Editor

Lei Ying, ECE, University of Michigan–Ann Arbor, Ann Arbor, USA

The series publishes short books on the design, analysis, and management of complex networked systems using tools from control, communications, learning, optimization, and stochastic analysis. Each Lecture is a self-contained presentation of one topic by a leading expert. The topics include learning, networks, and algorithms, and cover a broad spectrum of applications to networked systems including communication networks, data-center networks, social, and transportation networks.

Gauri Joshi

Optimization Algorithms for Distributed Machine Learning

 Springer

Gauri Joshi
Carnegie Mellon University
Pittsburgh, PA, USA

ISSN 2690-4306   ISSN 2690-4314   (electronic)
Synthesis Lectures on Learning, Networks, and Algorithms
ISBN 978-3-031-19069-8   ISBN 978-3-031-19067-4   (eBook)
https://doi.org/10.1007/978-3-031-19067-4

This Springer imprint is published by the registered company Springer Nature Switzerland AG
The registered company address is: Gewerbestrasse 11, 6330 Cham, Switzerland

To my students and collaborators for their dedicated work and insightful discussions without which this book would not have been possible

To my family for their unconditional support and encouragement

Preface

Stochastic gradient descent is the backbone of supervised machine learning training today. Classical SGD was designed to be run on a single computing node, and its error convergence with respect to the number of iterations has been extensively analyzed and improved in optimization and learning theory literature. However, due to the massive training datasets and models used today, running SGD at a single node can be prohibitively slow. This calls for distributed implementations of SGD, where gradient computation and aggregation are split across multiple worker nodes. Although parallelism boosts the amount of data processed per iteration, it exposes SGD to unpredictable node slowdown and communication delays stemming from variability in the computing infrastructure. Thus, there is a critical need to make distributed SGD fast, yet robust to system variability.

In this book, we will discuss state-of-the-art algorithms in large-scale machine learning that improve the scalability of distributed SGD via techniques such as asynchronous aggregation, local updates, quantization and decentralized consensus. These methods reduce the communication cost in several different ways—asynchronous aggregation allows overlap between communication and local computation, local updates reduce the communication frequency thus amortizing the communication delay across several iterations, quantization and sparsification methods reduce the per-iteration communication time, and decentralized consensus offers spatial communication reduction by allowing different nodes in a network topology to train models and average them with neighbors in parallel.

For each of the distributed SGD algorithms presented here, the book also provides an analysis of its convergence. However, unlike traditional optimization literature, we do not only focus on the error versus iterations convergence, or the iteration complexity. In distributed implementations, it is important to study the error versus wallclock time convergence because the wallclock time taken to complete each iteration is impacted by the synchronization and communication protocol. We model computation and communication delays as random variables and determine the expected wallclock runtime per iteration of the various distributed SGD algorithms presented in this book. By pairing this runtime analysis with the error convergence analysis, one can get a true comparison of the convergence speed of different algorithms. The book advocates a system-aware philosophy, which is cognizant of computation, synchronization and communication delays, toward the design and analysis of distributed machine learning algorithms.

This book would not have been possible without the wonderful research done by my students and collaborators. I thank them for helping me learn the material presented in this book. Our research was generously supported by several funding agencies including the National Science Foundations and research awards from IBM, Google and Meta. I was also inspired by the enthusiasm of the students who took my class on large-scale machine learning infrastructure over the past few years. Finally, I am immensely grateful to my family and friends for their constant support and encouragement.

Pittsburgh, PA, USA   Gauri Joshi
August 2022

Contents

1 Distributed Optimization in Machine Learning 1
  1.1 SGD in Supervised Machine Learning 1
      1.1.1 Training Data and Hypothesis 1
      1.1.2 Empirical Risk Minimization 2
      1.1.3 Gradient Descent 2
      1.1.4 Stochastic Gradient Descent 3
      1.1.5 Mini-batch SGD 3
      1.1.6 Linear Regression 4
      1.1.7 Logistic Regression 6
      1.1.8 Neural Networks 6
  1.2 Distributed Stochastic Gradient Descent 7
      1.2.1 The Parameter Server Framework 8
      1.2.2 The System-Aware Design Philosophy 8
  1.3 Scalable Distributed SGD Algorithms 9
      1.3.1 Straggler-Resilient and Asynchronous SGD 9
      1.3.2 Communication-Efficient Distributed SGD 10
      1.3.3 Decentralized SGD 10
  References 11

2 Calculus, Probability and Order Statistics Review 13
  2.1 Calculus and Linear Algebra 13
      2.1.1 Norms and Inner Products 13
      2.1.2 Lipschitz Continuity and Smoothness 14
      2.1.3 Strong Convexity 16
  2.2 Probability Review 17
      2.2.1 Random Variable 18
      2.2.2 Expectation and Variance 18
      2.2.3 Some Canonical Random Variables 19
      2.2.4 Bayes Rule and Conditional Probability 20

2.3   Order Statistics   22
    2.3.1   Order Statistics of the Exponential Distribution   22
    2.3.2   Order Statistics of the Uniform Distribution   24
    2.3.3   Asymptotic Distribution of Quantiles   24

3   Convergence of SGD and Variance-Reduced Variants   27
3.1   Gradient Descent (GD) Convergence   27
    3.1.1   Effect of Learning Rate and Other Parameters   28
    3.1.2   Iteration Complexity   29
3.2   Convergence Analysis of Mini-batch SGD   29
    3.2.1   Effect of Learning Rate and Mini-batch Size   32
    3.2.2   Iteration Complexity   32
    3.2.3   Non-convex Objectives   33
3.3   Variance-Reduced SGD Variants   34
    3.3.1   Dynamic Mini-batch Size Schedule   34
    3.3.2   Stochastic Average Gradient (SAG)   35
    3.3.3   Stochastic Variance Reduced Gradient (SVRG)   36
References   38

4   Synchronous SGD and Straggler-Resilient Variants   39
4.1   Parameter Server Framework   39
4.2   Distributed Synchronous SGD Algorithm   40
4.3   Convergence Analysis   41
    4.3.1   Iteration Complexity   42
4.4   Runtime per Iteration   42
    4.4.1   Gradient Computation and Communication Time   43
    4.4.2   Expected Runtime per Iteration   43
    4.4.3   Error Versus Runtime Convergence   45
4.5   Straggler-Resilient Variants   45
    4.5.1 K-Synchronous SGD   46
    4.5.2 K-Batch-Synchronous SGD   47
References   49

5   Asynchronous SGD and Staleness-Reduced Variants   51
5.1   The Asynchronous SGD Algorithm   51
    5.1.1   Comparison with Synchronous SGD   52
5.2   Runtime Analysis   53
    5.2.1   Runtime Speed-Up Compared to Synchronous SGD   53
5.3   Convergence Analysis   54
    5.3.1   Implications of the Asynchronous SGD Convergence Bound   57
5.4   Staleness-Reduced Variants of Asynchronous SGD   57
    5.4.1 K-Asynchronous SGD   58

5.4.2 K-Batch-Asynchronous SGD 60
5.5 Adaptive Methods to Improve the Error-Runtime Trade-Off 61
5.5.1 Adaptive Synchronization 61
5.5.2 Adaptive Learning Rate Schedule to Compensate Staleness 63
5.6 HogWild and Lock-Free Parallelism 64
References 66

6 Local-Update and Overlap SGD 67
6.1 Local-Update SGD Algorithm 67
6.1.1 Convergence Analysis 69
6.1.2 Runtime Analysis 76
6.1.3 Adaptive Communication 79
6.2 Elastic and Overlap SGD 84
6.2.1 Elastic Averaging SGD 85
6.2.2 Overlap Local SGD 87
References 91

7 Quantized and Sparsified Distributed SGD 93
7.1 Quantized SGD 93
7.1.1 Uniform Stochastic Quantization 94
7.1.2 Convergence Analysis 95
7.1.3 Runtime Analysis 98
7.1.4 Adaptive Quantization 100
7.2 Sparsified SGD 101
7.2.1 Rand-k Sparsification 102
7.2.2 Top-k Sparsification 102
7.2.3 Rand-k Sparsified Distributed SGD 103
7.2.4 Error Feedback in Sparsified SGD 104
References 106

8 Decentralized SGD and Its Variants 107
8.1 Network Topology and Graph Notation 107
8.1.1 Adjacency Matrix 108
8.1.2 Laplacian Matrix 108
8.1.3 Mixing Matrix 109
8.2 Decentralized SGD 109
8.2.1 The Algorithm 110
8.2.2 Variants of Decentralized SGD 111
8.3 Error Convergence Analysis 112
8.3.1 Assumptions 113
8.3.2 Convergence Analysis of Decentralized SGD 113

        8.3.3 Convergence Analysis of Decentralized Local-Update SGD 117
    8.4 Runtime Analysis 118
    References 121
9 **Beyond Distributed Training in the Cloud** 123
    References 126

Acronyms and Symbols

$\mathcal{D}$	Training dataset
N	Number of samples in the training dataset
$\mathbf{x}$	Feature vector corresponding to a training example
y	Label corresponding to a training example
$\mathbf{w}$	Parameter vector of the machine learning model
d	Dimension of the model parameter vector $\mathbf{w}$
GD	Gradient descent
SGD	Stochastic gradient descent
b	Mini-batch size
η	Learning rate or step size
m	Number of worker nodes
L	Lipschitz smoothness parameter
c	Strong convexity parameter
σ^2	Stochastic gradient variance bound
τ	Number of local updates at each worker
$\mathbf{M}$	Mixing matrix

Distributed Optimization in Machine Learning

1

Machine learning is revolutionizing data-driven decision-making in a myriad applications including search and recommendations, self-driving cars, robotics, and medical diagnosis. And stochastic gradient descent (SGD) is the most common algorithm used for training such machine learning models. Due to the massive size of training datasets in modern applications, SGD is typically implemented in a distributed fashion, where the task of computing gradients is split across multiple computing nodes. The goal of this book is to introduce you to the state-of-the-art distributed optimization algorithms used in machine learning. For each distributed SGD algorithm, we will take a deep dive into analyzing this error versus runtime performance trade-off. These analyses will equip you with insights on designing algorithms that are best suited to the communication and computation constraints of the systems infrastructure.

We begin this chapter by introducing how SGD is employed in supervised machine learning. Next, we will present the motivation for distributing the implementation of SGD across multiple computing nodes. In designing distributed SGD algorithms, this book advocates a system-aware philosophy. Unlike classic optimization theory which focuses on the sample or iteration complexity of SGD algorithms, the system-aware philosophy is cognizant of communication and synchronization delays in the underlying computing infrastructure. In the rest of the book, we will deep dive into various scalable distributed SGD algorithms, which are summarized in the last section of this chapter.

1.1   SGD in Supervised Machine Learning

1.1.1   Training Data and Hypothesis

Supervised learning is a class of machine learning, where the goal is to learn a prediction function or a hypothesis $h(\mathbf{x}) : \mathcal{X} \rightarrow \mathcal{Y}$ that maps an input *feature* vector $\mathbf{x} \in \mathcal{X}$ to an output *label* $y \in \mathcal{Y}$. For example, if our goal is to identify spam emails, then the feature vector $\mathbf{x}$

© The Author(s), under exclusive license to Springer Nature Switzerland AG 2023
G. Joshi, *Optimization Algorithms for Distributed Machine Learning*,
Synthesis Lectures on Learning, Networks, and Algorithms,
https://doi.org/10.1007/978-3-031-19067-4_1

contains the number of occurrences of suspicious words in an email, and the function h maps $\mathbf{x}$ to a binary label $y \in \{0, 1\}$ such that $y = 1$ indicates that the email is spam. How do we learn the hypothesis h that will achieve the goal of correctly mapping an input feature to an output label? We use a *training dataset*, consisting of known feature-label pairs $(\mathbf{x}_1, y_1), (\mathbf{x}_2, y_2), \ldots, (\mathbf{x}_N, y_N)$. The training dataset *supervises* the hypothesis h in learning to predict the label y' of an unseen feature vector $\mathbf{x}'$.

1.1.2 Empirical Risk Minimization

In order to measure how good $h(\mathbf{x})$ is at predicting the correct label, we define the sample loss $\ell(h(\mathbf{x}_i), y_i)$ for the training sample $(\mathbf{x}_n, y_n), n \in \{1, \ldots, N\}$. For example, if our goal is to perform regression and match $h(\mathbf{x})$ to the target value y, then a common choice of the sample loss $\ell(h(\mathbf{x}_i), y_i) = (h(\mathbf{x}) - y)^2$, the square loss function. The training phase of machine learning seeks to minimize the average loss over the entire training dataset, which is also referred to as the empirical risk objective function [1, 2]:

$$F(h) = \frac{1}{N} \sum_{n=1}^{N} \ell(h(\mathbf{x}_n), y_n) \qquad (1.1)$$

The objective function defined in (1.1) is referred to as the *empirical* risk because the training dataset $(\mathbf{x}_1, y_1), (\mathbf{x}_2, y_2), \ldots, (\mathbf{x}_N, y_N)$ is a finite sample of the joint probability distribution $P(\mathbf{x}, y)$ of the features and the labels. The goal of supervised training is to find the hypothesis h^* from within a hypothesis class $\mathcal{H}$ that minimizes the empirical risk, that is,

$$h^* = \arg \min_{h \in \mathcal{H}} F(h). \qquad (1.2)$$

1.1.3 Gradient Descent

For simple hypothesis classes $\mathcal{H}$ and sample losses $\ell(\cdot)$ it may be possible to directly solve for h^*. However, in general, we need to employ an iterative algorithm such as gradient descent (GD) to solve the optimization problem in (1.2). Consider a hypothesis class $\mathcal{H}$ for which each hypothesis h is specified by a d-dimensional parameter vector $\mathbf{w}$, and the corresponding empirical risk objective $F(h)$ can be written as $F(\mathbf{w})$. Gradient descent starts with a randomly initialized set of parameters $\mathbf{w}_0$. Then in each iteration, it computes the gradient $\nabla F(h)$ of the loss function at $\mathbf{w}$ and takes a small step in the direction of steepest descent as follows:

$$\mathbf{w}_{t+1} = \mathbf{w}_t - \eta \nabla F(\mathbf{w}_t) \tag{1.3}$$

$$= \mathbf{w}_t - \frac{\eta}{N} \sum_{n=1}^{N} \nabla \ell(h(\mathbf{x}_n), y_n)) \tag{1.4}$$

where η is the step size or the learning rate. Since each iteration of GD uses all the N samples in training dataset, it is sometimes referred to batch GD or full-batch GD.

1.1.4 Stochastic Gradient Descent

Observe that the computational complexity of each iteration of gradient descent is $O(Nd)$, which can be prohibitively expensive for a large training dataset size N and for high-dimensional $\mathbf{w}$ (that is, large d). A computationally efficient alternative is to use *stochastic gradient descent*, first proposed in [3], where the gradient $\nabla F(\mathbf{w}_t)$ is replaced by a noisy estimate $\nabla \ell(h(\mathbf{x}_n), y_n))$ computed using a single training example $(\mathbf{x}_n, y_n)$. In each iteration of stochastic gradient descent,

$$\mathbf{w}_{t+1} = \mathbf{w}_t - \eta \nabla \ell(h(\mathbf{x}_n), y_n)) \tag{1.5}$$

$$= \mathbf{w}_t - \eta \nabla \ell(\mathbf{w}_t, \xi_t) \tag{1.6}$$

The training example $(\mathbf{x}_n, y_n)$ is sampled uniformly at random with replacement from the training dataset. The symbol ξ_t is often used to denote the training example sampled in the $t+1$-th iteration when going from $\mathbf{w}_t$ to $\mathbf{w}_{t+1}$, and thus the stochastic gradient is written as $g(\mathbf{w}_t) = \nabla \ell(\mathbf{w}_t, \xi_t)$. Observe that the computational complexity of each iteration of SGD is $O(d)$, which is independent of the size of the dataset. For large datasets it is significantly smaller than the $O(Nd)$ complexity of gradient descent.

1.1.5 Mini-batch SGD

Since we are using a noisy estimate of the gradient, SGD typically takes more iterations to converge close to the optimal parameters $\mathbf{w}^*$ that minimize the empirical risk objective function. For example, Fig. 1.1 shows the trajectory of GD and SGD for a convex loss function. To achieve a middle ground between these two extremes, most practical implementations use an algorithm called *mini-batch SGD*, where in each iteration we sample a batch of b training example uniformly at random with replacement, and then use their averaged sample loss gradient $g(\mathbf{w}_t) = \sum_{l=1}^{b} \nabla \ell(\mathbf{w}_t, \xi_{t,l})/b$ in place of $\nabla \ell(\mathbf{w}_t, \xi_t)$ in (1.6) to update the model parameters $\mathbf{w}$:

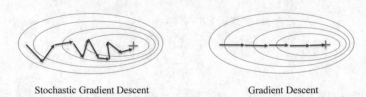

Stochastic Gradient Descent   Gradient Descent

Fig. 1.1 Illustration of the convergence of stochastic gradient descent (SGD) versus gradient descent (GD). SGD takes a noisy path towards the minimum because the gradients are noisy estimates of the full gradient

$$\mathbf{w}_{t+1} = \mathbf{w}_t - \frac{\eta}{b} \sum_{l=1}^{b} \nabla \ell(\mathbf{w}_t, \xi_{t,l}) \tag{1.7}$$

Observe that mini-batch SGD reduces to GD if we set $b = N$ and it reduces to SGD if we set $b = 1$. For a suitable intermediate value of b, mini-batch SGD is more computationally-efficient than GD, but significantly less noisy than SGD.

In Chap. 3 we will formally study the convergence of gradient descent and stochastic gradient descent to quantify the number of iterations T required until the distance between expected value of $\mathbf{w}_T$ and $\mathbf{w}^*$ reaches below an ϵ target error. In Chap. 2 we review concepts from probability, linear algebra and calculus that are necessary to understand the proofs presented in Chap. 3.

Although the convergence to the optimal $\mathbf{w}^*$ is guaranteed only for convex objectives, mini-batch SGD has been shown to perform well even on non-convex loss surfaces due to its ability to escape saddle points and local minima [4–7]. As a result, it is the dominant algorithm used in machine learning.

Several works such as [8–15] focus on accelerating the convergence of mini-batch SGD in terms of error versus the number of iterations. In Chap. 3 we will study some of these variants such as variance-reduction techniques in SGD. In Sects. 1.1.6, 1.1.7 and 1.1.8 below we review three examples of SGD in machine learning, namely linear regression, logistic regression and neural networks. For a more detailed treatment, please refer an introductory machine learning textbook [2].

1.1.6   Linear Regression

Linear regression is the simplest form of machine learning, where the hypothesis class $\mathcal{H}$ is restricted to linear mappings from the features $\mathbf{x} = [x_1, \ldots, x_d]^\top \in \mathbb{R}^d$ to the label (also referred as the target) $y \in \mathbb{R}$. Thus, a hypothesis h is given by:

$$h(\mathbf{x}) = w_0 + w_1 x_1 + \cdots w_d x_d \tag{1.8}$$

$$= [w_0, \ldots, w_d] \begin{bmatrix} 1 \\ x_1 \\ x_2 \\ \vdots \\ x_d \end{bmatrix} \tag{1.9}$$

$$= \mathbf{w}^\top \mathbf{x}' \tag{1.10}$$

where $\mathbf{w} = [w_0, w_1, \ldots, w_d]^\top$ is the model parameter vector and w_0 is often referred to as the bias. The vector $\mathbf{x}'^\top = [1 \ \ \mathbf{x}^\top]$ is the augmented feature vector, which includes a 1 corresponding to the bias parameter w_0.

The performance of the linear hypothesis function for a data sample is measured in terms of the square sample loss function:

$$\ell(h(\mathbf{x}_i), y_i) = (y_i - h(\mathbf{x}))^2. \tag{1.11}$$

For a training dataset $(\mathbf{x}_1, y_1), (\mathbf{x}_2, y_2), \ldots, (\mathbf{x}_N, y_N)$, the empirical risk objective function becomes the least mean squares function given by:

$$F(\mathbf{w}) = \frac{1}{N} \sum_{n=1}^{N} (y_i - \mathbf{w}^\top \mathbf{x}')^2 \tag{1.12}$$

$$= \frac{1}{N} \|\mathbf{y} - \mathbf{X}\mathbf{w}\|^2 \tag{1.13}$$

where

$$\mathbf{X} = \begin{bmatrix} \mathbf{x}_1'^\top \\ \mathbf{x}_2'^\top \\ \vdots \\ \mathbf{x}_N'^\top \end{bmatrix} \quad \text{and} \quad \mathbf{y} = \begin{bmatrix} y_1 \\ y_2 \\ \vdots \\ y_N \end{bmatrix} \tag{1.14}$$

which are referred to as the design matrix and the target vector respectively.

Due to the simple form of the empirical risk objective function in (1.13), we can directly get a closed-form expression for the model parameters $\mathbf{w}^*$ that minimize $F(\mathbf{w})$:

$$\mathbf{w}^* = \min_{\mathbf{w} \in \mathbb{R}^{d+1}} \frac{1}{N} \|\mathbf{y} - \mathbf{X}\mathbf{w}\|^2 = (\mathbf{X}^\top \mathbf{X})^{-1} \mathbf{X}^\top \mathbf{y}, \tag{1.15}$$

referred to as the least mean squares solution.

For a training dataset with a large number of training examples N, and a high-dimensional model (large d) evaluating the least mean squares solution $\mathbf{w}^*$ can be computationally expensive. The dominant terms in the computation complexity are the $O(Nd)$ operations to compute the matrix product $\mathbf{X}^\top \mathbf{X}$, and the $O(d^3)$ operations to invert $\mathbf{X}^\top \mathbf{X}$. Gradient descent and its variants described in (1.4), (1.6) and (1.7) above are computationally-efficient alternatives to find the least squares solution $\mathbf{w}^*$.

1.1.7   Logistic Regression

When the labels y take categorical values rather than real numbered values, for example, binary values 0 or 1, the squared loss function $(y - h(\mathbf{x}))^2$ defined in (1.11) is not suitable to measure the error in *classifying* feature vectors $\mathbf{x}$ into the categories 0 or 1. And instead of a linear hypothesis $h(\mathbf{x}) = \mathbf{w}^\top \mathbf{x}$, we need a non-linear function that swiftly transitions from 0 to 1. Logistic regression uses the sigmoid function $\sigma(\mathbf{w}^\top \mathbf{x}) = 1/(1 + e^{-\mathbf{w}^\top \mathbf{x}})$ to denote the probability that the predicted label y is 0 or 1. If $\mathbf{w}^\top \mathbf{x} < 0$, then the predicted label is assigned as 0, and if $\mathbf{w}^\top \mathbf{x} > 0$ the predicted label is set to 1.

In order to train the parameters $\mathbf{w}$ that determine the decision boundary, we minimize the following cross entropy loss function for training sample $(\mathbf{x}_n, y_n)$:

$$\ell(\sigma(\mathbf{w}^\top \mathbf{x}_n), y_n) = -\left(y_n \log \sigma(\mathbf{w}^\top \mathbf{x}_n) + (1 - y_n) \log(1 - \sigma(\mathbf{w}^\top \mathbf{x}_n)) \right) \qquad (1.16)$$

It represents the negative log of the likelihood $\Pr(y_n | \mathbf{x}_n; \mathbf{w}) = \sigma(\mathbf{w}^\top \mathbf{x}_n)^{y_n} (1 - \sigma(\mathbf{w}^\top \mathbf{x}_n))^{1-y_n}$ of sample. For a training dataset $(\mathbf{x}_1, y_1), (\mathbf{x}_2, y_2), \ldots, (\mathbf{x}_N, y_N)$, the empirical risk objective function is:

$$F(\mathbf{w}) = -\frac{1}{N} \sum_{n=1}^{N} \left(y_n \log \sigma(\mathbf{w}^\top \mathbf{x}_n) + (1 - y_n) \log(1 - \sigma(\mathbf{w}^\top \mathbf{x}_n)) \right) \qquad (1.17)$$

Unlike the least-squares solution for linear regression, we cannot get a closed-form expression for the optimal parameter vector $\mathbf{w}^*$ that minimizes the $F(\mathbf{w})$. Thus, we have to resort to using iterative methods such as gradient descent to train the parameters. The (batch) gradient descent update rule for logistic regression is given by:

$$\mathbf{w}_{t+1} = \mathbf{w}_t - \frac{\eta}{N} \sum_{n=1}^{N} (\sigma(\mathbf{w}^\top \mathbf{x}_n) - y_n) \qquad (1.18)$$

using the property that the derivation of sigmoid, $\sigma'(a) = \sigma(a)(1 - \sigma(a))$, for any $a \in \mathbb{R}$.

1.1.8   Neural Networks

Logistic regression enables us to learn *linear* decision boundaries, given by $\mathbf{w}^\top \mathbf{x} = 0$. Neural networks are a flexible and efficient way to learn more complex hypothesis functions $h(\mathbf{x})$ mapping a feature vector $\mathbf{x}$ to labels y. Neural networks consist of computation units called

Fig. 1.2 Neural network with one hidden layer

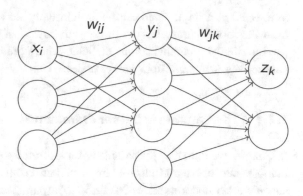

neurons, which are organized into layers, as illustrated in Fig. 1.2. Each neuron applies a non-linear activation function $g(\ldots)$ to the linear combination of its inputs. For example, the output y_j of the hidden layer neuron shown in Fig. 1.2 is $y_j = g(\sum_i w_{ij} x_i)$. The objective function $F(\mathbf{w})$ measures the error between the output z_k of the last layer and the target or label y_k.

Stochastic gradient descent is the dominant algorithm used to minimize this objective function and tune the weight of each of the connections of the neural networks. Due to the layer-wise structure of the network, the gradients of the loss $F(\mathbf{w})$ with respect to each weight w_{ij} can be computed in a recursive fashion. This recursive algorithm to compute the gradients in an efficient manner is called the backpropagation algorithm, first introduced in [16]. Since the objective function of neural networks is highly non-convex, SGD is not guaranteed to converge to the global minimum. However, in practice, convergence to a local minimum is often sufficient to achieve high training and test accuracy.

Since the focus of this book is on the convergence of distributed SGD algorithm, we will abstract the exact form of the objective function $F(\mathbf{w})$. It could be the linear regression loss, logistic regression loss, a neural network loss function, or the loss function of any other machine learning model.

1.2   Distributed Stochastic Gradient Descent

Classical SGD was designed to be run on a single computing node, and its error-convergence with respect to the number of iterations has been extensively analyzed and improved in optimization and learning theory literature. However, in modern machine learning applications, the massive training data-sets and deep neural network architectures are massive. For example, the widely used ImageNet dataset [17] used to train image classification models $N \sim 1.3$ million images across 1000 classes. And the commonly used neural network architecture called ResNet-50 [18] for image classification has $d = 23$ million trainable parameters. Since SGD is an inherently sequential algorithm, for such large datasets and models, running SGD at a single node can be prohibitively slow. It may take days or even weeks to train the model parameters **w**. In order to cut down the training time, it is imperative to use distributed implementations of SGD, where gradient computation and aggregation is parallelized across multiple worker nodes.

1.2.1 The Parameter Server Framework

In Chap. 4, we introduce the parameter server framework that is used to run distributed SGD in most state-of-the-art industrial implementations. It consists of a central parameter server and m worker nodes, as shown in Fig. 5.1. The parameter server stores and updates the model parameters **w**. The training dataset is shuffled and split equally into partitions $\mathcal{D}_1, ..., \mathcal{D}_m$ which are stored at the m workers.

The most common distributed SGD algorithm in the parameter server framework is called *synchronous SGD*. In the $t + 1$-th iteration of synchronous SGD, each of the m worker nodes reads the current version of $\mathbf{w}_t$ from the parameter server and computes a gradient $g(\mathbf{w}_t, \xi_i)$ using a mini-batch (denoted by ξ_i) of b samples drawn from its local dataset partition $\mathcal{D}_i$. The parameter server then collects the gradients and updates the model **w** as per the following update rule:

$$\mathbf{w}_{t+1} = \mathbf{w}_t - \frac{\eta}{m} \sum_{i=1}^{m} g(\mathbf{w}_t, \xi_i) \tag{1.19}$$

$$= \mathbf{w}_t - \frac{\eta}{mb} \sum_{i=1}^{m} \sum_{l=1}^{b} \nabla \ell(\mathbf{w}_t, \xi_{t,l}) \tag{1.20}$$

The synchronous SGD update rule defined above is, in fact, equivalent to mini-batch SGD with a mini-batch size of bm instead of b. By utilizing m worker nodes that compute gradients in parallel, we are able to process m times more data per iteration.

1.2.2 The System-Aware Design Philosophy

The main benefit of using more workers is that it reduces the noise in the averaged gradient used for updating **w** in every iteration and thus improves the error versus iterations convergence. The number of iterations required to achieve a target error, also referred to as the iteration complexity has been extensively studied in optimization theory and many SGD variants have been proposed to accelerated the convergence. However, when SGD is run in a distributed setting, it is not sufficient to focus on the error versus iterations convergence because the wall-clock time spent per iteration depends on the synchronization and communication delays stemming from the underlying computing infrastructure.

In this book, we emphasize a system-aware design philosophy that considers the true convergence speed of distributed SGD with respect to the wallclock runtime. It is a product of two factors: (1) the error in the trained model versus the number of iterations, and (2) the number of iterations completed per second. Traditional single-node SGD analysis focuses on optimizing the first factor, because the second factor is generally a constant when SGD is run on a single dedicated server. In distributed SGD, which is often run on shared cloud infrastructure, the second factor depends on several aspects such as the gradient computation and communication delays at the worker nodes, and the protocol (synchronous, asynchronous or periodic) used to aggregate their gradients. Hence, in order to achieve the fastest convergence speed we need: (1) optimization techniques (e.g. variable learning rate) to maximize the error-convergence rate with respect to iterations, and (2) scheduling techniques (e.g. straggler mitigation) to maximize the number of iterations completed per second. These directions are inter-dependent and need to be explored together rather than in isolation. While many works have advanced the first direction, the second is less explored from a theoretical point of view, and the juxtaposition of both is an unexplored problem.

In Chap. 4, we adopt this philosophy to analyze the error versus runtime convergence of synchronous SGD. We quantify how the number of workers m affects the convergence speed by combining (1) the error-versus-iterations convergence analysis of mini-batch SGD derived in Chap. 3 and (2) the analysis of how m and the gradient computation delay distribution at each workers affects the expected wallclock runtime per iteration.

1.3   Scalable Distributed SGD Algorithms

Synchronous SGD, the simplest distributed SGD algorithm does not scale well to a large number of worker nodes due to synchronization and communication delays and the bandwidth limitations at the central parameter server. In this book, we will study several scalable variants of distributed SGD that are robust to such inherent variability and constraints of the computing infrastructure.

1.3.1   Straggler-Resilient and Asynchronous SGD

In each iteration of synchronous SGD, the parameter server has to wait for *all* the m to send back gradients before it can update the model **w** and proceed to the next iteration. Worker nodes being cloud servers, are suspectible to unpredictable slowdown or failure due to various reasons such as background workload, outages, etc. Such node straggling is the norm rather than the exception in data center computing [19]. As the number of workers m increases, even a small probability of straggling can cause an order-of-magnitude increase the expected runtime per iteration. In Chap. 5 we propose straggler-resilient variants of synchronous SGD that are robust to straggling workers. These variants span different points on the trade-off between iterations complexity and runtime per iteration. Relaxing the synchronization barrier of synchronous SGD can reduce the runtime per iteration but it may increase the iteration complexity, that is, the number of iterations required to reach a target training loss. We present convergence analysis and runtime analysis to quantify this trade-off.

1.3.2 Communication-Efficient Distributed SGD

Besides the time taken by workers to compute its mini-batch gradient, the runtime per iteration also includes the communication delay spent in sending the gradients to the parameter server and receiving the updated model parameter. When the model **w** is high-dimensional or when the communication link has high latency, this delay can dominate the runtime per iteration. This is especially true in emerging distributed machine learning framework such as federated learning where the worker nodes are edge devices such as cell phones, which have unreliable and low-bandwidth wireless links to connect with the cloud, where the parameter server is present. Therefore there is a critical need to design communication-efficient distributed SGD algorithms, going beyond synchronous SGD which requires communication after every iteration.

In this book we will study two orthogonal ways to reduce communication delay. The first approach is to reduce the communication frequency by using an algorithm called *local-update SGD*, where the workers perform multiple local SGD iterations instead of just computing gradients, and the resulting locally trained models are averaged by the parameter server. In Chap. 6 we will study the convergence and runtime properties of local SGD and its variant elastic-averaging SGD. Elastic-averaging SGD further improves the scalability of local-update SGD by facilitating the overlap of communication and computation. The second approach to make distributed SGD more communication-efficient is to reduce the number of bits transmitted from the workers to the server by quantizing or sparsifying the gradient or model parameters. In Chap. 7 we will study quantization and sparsification techniques. Since the compression is lossy, it can have an adverse effect of the error at convergence. We will study these convergence and runtime trade-offs in Chap. 7.

1.3.3 Decentralized SGD

As the number of worker nodes m grows, it may be prohibitively expensive to have a single central parameter server to aggregate the gradients or model updates and maintain the latest version of the model parameters **w**. Instead, training can be performed in a decentralized fashion where there is an arbitrary network topology connecting the workers, as shown in Fig. 1.3. Each worker makes local updates to the model parameters based on its local datasets and then averages the updates with its neighbors. Eventually, as long as the network topology is connected, the updates from a worker will reach all other workers. In Chap. 8

Fig. 1.3 Decentralized SGD topology consisting of 8 nodes

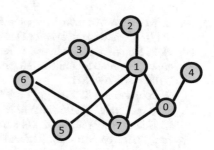

we will study decentralized SGD algorithms and their convergence analysis. The number of iterations required for the model to converge to its optimal value depends on the number of worker nodes, the topology connecting them and the inter-node communication delays. Decentralized SGD has many applications such as multi-agent networks of sensors or IoT devices and cross-silo federated learning.

Summary

Stochastic gradient descent (SGD) is at the core of state-of-the-art supervised learning, and due to large datasets and models used today, it is imperative to implement it in a distributed manner. Thus, speeding-up distributed SGD is arguably the single most impactful and transformative problem in the field of machine learning. This book takes a system-aware approach towards designing distributed SGD algorithms, which is cognizant of the synchronization and communication delays in the computing infrastructure. In this chapter, we outlined the various scalable distributed SGD algorithms that we will study in this book.

References

1. S. Boyd and L. Vandenberghe, *Convex optimization.* Cambridge university press, 2004.
2. S. Shalev-Shwartz and S. Ben-David, *Understanding Machine Learning: From Theory to Algorithms.* New York, NY, USA: Cambridge University Press, 2014.
3. H. Robbins and S. Monro, "A stochastic approximation method," *The annals of mathematical statistics*, pp. 400–407, 1951.
4. C. Zhang, S. Bengio, M. Hardt, B. Recht, and O. Vinyals, "Understanding deep learning requires rethinking generalization," 2017. [Online]. Available: https://arxiv.org/abs/1611.03530
5. B. Neyshabur, R. Tomioka, R. Salakhutdinov, and N. Srebro, "Geometry of optimization and implicit regularization in deep learning," *CoRR*, vol. abs/1705.03071, 2017. [Online]. Available: http://arxiv.org/abs/1705.03071
6. P. Chaudhari and S. Soatto, "Stochastic gradient descent performs variational inference, converges to limit cycles for deep networks," *CoRR*, vol. abs/1710.11029, 2017. [Online]. Available: http://arxiv.org/abs/1710.11029
7. R. Shwartz-Ziv and N. Tishby, "Opening the black box of deep neural networks via information," *CoRR*, vol. abs/1703.00810, 2017. [Online]. Available: http://arxiv.org/abs/1703.00810
8. B. T. Polyak, "Some methods of speeding up the convergence of iteration methods," *USSR Computational Mathematics and Mathematical Physics*, vol. 4, no. 5, pp. 1–17, 1964.

9. Y. Nesterov, "A method of solving a convex programming problem with convergence rate $o(1/k^2)$," *Soviet Mathematics Doklady*, vol. 27, no. 5, pp. 372–376, 1983.

10. N. L. Roux, M. Schmidt, and F. R. Bach, "A stochastic gradient method with an exponential convergence rate for finite training sets," in *Advances in Neural Information Processing Systems*, 2012, pp. 2663–2671.

11. R. Johnson and T. Zhang, "Accelerating stochastic gradient descent using predictive variance reduction," in *Advances in Neural Information Processing Systems 26*, C. J. C. Burges, L. Bottou, M. Welling, Z. Ghahramani, and K. Q. Weinberger, Eds.   Curran Associates, Inc., 2013, pp. 315–323. [Online]. Available: http://papers.nips.cc/paper/4937-accelerating-stochastic-gradient-descent-using-predictive-variance-reduction.pdf

12. Z. Allen-Zhu, "Katyusha: The first direct acceleration of stochastic gradient methods," in *Proceedings of Symposium on Theory of Computing (STOC)*, 2017, pp. 1200–1205.

13. L. Nguyen, J. Liu, K. Scheinberg, and M. Takáč, "Sarah: A novel method for machine learning problems using stochastic recursive gradient," *arXiv preprint* arXiv:1703.00102, 2017.

14. J. Duchi, E. Hazan, and Y. Singer, "Adaptive Subgradient Methods for Online Learning and Stochastic Optimization," *Journal of Machine Learning Research*, vol. 12, pp. 2121–2159, 2011.

15. D. P. Kingma and J. Ba, "Adam: A method for stochastic optimization," *International Conference on Learning Representations (ICLR)*, 2015.

16. D. E. Rumelhart, G. E. Hinton, and R. J. Williams, "Learning representations by back-propagating errors," *Nature*, vol. 323, pp. 533–536, 1986.

17. O. Russakovsky, J. Deng, H. Su, J. Krause, S. Satheesh, S. Ma, Z. Huang, A. Karpathy, A. Khosla, M. Bernstein, A. C. Berg, and L. Fei-Fei, "Imagenet large scale visual recognition challenge," *International Journal of Computer Vision*, vol. 115, no. 3, pp. 211–252, 2015.

18. K. He, X. Zhang, S. Ren, and J. Sun, "Deep residual learning for image recognition," in *Proceedings of the IEEE conference on computer vision and pattern recognition*, 2016, pp. 770–778.

19. J. Dean and L. A. Barroso, "The tail at scale," *Communications of the ACM*, vol. 56, no. 2, pp. 74–80, 2013.

Calculus, Probability and Order Statistics Review

2

In this chapter, we will begin by reviewing some concepts from calculus and probability that are essential to understanding the error and runtime analyses of various distributed SGD algorithms that we will study in the rest of the book.

2.1 Calculus and Linear Algebra

2.1.1 Norms and Inner Products

In machine learning problems, we quantify the quality of a hypothesis $h(\mathbf{x})$ by comparing it with the target y. However, $h(\mathbf{x})$ and $\mathbf{y}$ can be vectors. For example, in binary classification problems, the two elements of $h\mathbf{x}$ correspond to the probability or confidence with which the h function predicts the label as 0 and 1 respectively. The y vector is constructed using one-hot encoding where the i-th element is 1 if the true class is i, and otherwise it is zero. The difference $h(\mathbf{x}) - y$ between the predicted and the true labels is also a vector and it cannot be directly used to quantify the error.

Norms are functions that map a vector to a scalar value, and thus by taking a norm of $h(\mathbf{x}) - \mathbf{y}$ we can concretely quantify the error in the prediction $h\mathbf{x}$. The norm of a vector is defined as follows.

Definition 2.1 (*Vector norm*) A vector norm is a function that maps a vector $\mathbf{a} \in \mathbb{R}^d$ to a scalar $\|\mathbf{a}\|$ such that it satisfies the following conditions:

1. Non-negativity: $\|\mathbf{a}\| \geq 0$ and $\|\mathbf{a}\| = 0 \iff \mathbf{a} = 0$
2. Scaling: $\|c\mathbf{a}\| = |c| \, \|\mathbf{a}\|$ for any scalar $c \in \mathbb{R}$
3. Triangle Inequality: $\|\mathbf{a} + \mathbf{b}\| \leq \|\mathbf{a}\| + \|\mathbf{b}\|$

© The Author(s), under exclusive license to Springer Nature Switzerland AG 2023
G. Joshi, *Optimization Algorithms for Distributed Machine Learning*,
Synthesis Lectures on Learning, Networks, and Algorithms,
https://doi.org/10.1007/978-3-031-19067-4_2

Some commonly used vector norms are the ℓ_1 and ℓ_2 norms, $\|\mathbf{a}\|_1 = \sum_{i=1}^{d} |a_i|$ and $\|\mathbf{a}\|_2 = \sqrt{\sum_{i=1}^{d} a_i^2}$ respectively.

Another concept from linear algebra that we will use in this book is that of inner products of two vectors. While norms offer a way to measure the length of a vector by mapping it to a scalar quantity, inner products associate a pair of vectors with a scalar quantity. Inner products allow us a to measure the similarity of two vectors, and they are defined as follows.

Definition 2.2 (*Inner product*) The inner product of two vectors $\mathbf{a}_1, \mathbf{a}_2 \in \mathbb{R}^d$ is denoted by $\langle \mathbf{a}_1, \mathbf{a}_2 \rangle$ and it has to satisfy the following conditions:

1. $\langle c\mathbf{a}_1, \mathbf{a}_2 \rangle = c \langle \mathbf{a}_1, \mathbf{a}_2 \rangle$ for any scalar $c \in \mathbb{R}$
2. $\langle \mathbf{a}_1, \mathbf{a}_2 + \mathbf{a}_3 \rangle = \langle \mathbf{a}_1, \mathbf{a}_2 \rangle + \langle \mathbf{a}_1, \mathbf{a}_3 \rangle$
3. $\langle \mathbf{a}_1, \mathbf{a}_2 \rangle = \langle \mathbf{a}_2, \mathbf{a}_1 \rangle$
4. $\langle \mathbf{a}, \mathbf{a} \rangle > 0$ if $\mathbf{a} > 0$

The dot product is a special case of inner products where we take the sum of the element-wise product of the two vectors, that is, $\mathbf{a}_1 \cdot \mathbf{a}_2 = \sum_{i=1}^{d} a_{1,i} a_{2,i}$. The Euclidean or ℓ_2 norm is also a special case of inner product, where we take the dot product of a vector with itself, $\|\mathbf{a}\|_2^2 = \mathbf{a} \cdot \mathbf{a}$.

2.1.2   Lipschitz Continuity and Smoothness

The next concept that we will repeatedly use in this book is that of Lipschitz continuity and smoothness. Lipschitz continuity is a stronger form of continuity that restricts how quickly a function F can change.

Definition 2.3 (*Lipschitz continuity*) A function $F(\mathbf{x}) : \mathbb{R}^d \to \mathbb{R}$ is said to be K-Lipschitz continuous for some positive real constant $K > 0$ if

$$|F(\mathbf{x}_1) - F(\mathbf{x}_2)| \leq K |\mathbf{x}_1 - \mathbf{x}_2| \quad \text{for all } \mathbf{x}_1, \mathbf{x}_2 \in \mathbb{R}^d \qquad (2.1)$$

For a 1-dimensional function, Lipschitz continuity means that if you draw two lines with slopes K and $-K$ out of any point $(x, F(x))$ then the whole function lies between the two lines, as illustrated in Fig. 2.1.

Instead of restricting how quickly a function F can change, if we restrict how quickly its gradient will change, then we get the concept of Lipschitz smoothness that is formally defined as follows.

Definition 2.4 (*Lipschitz smoothness*) A function $F(\mathbf{x}) : \mathbb{R}^d \to \mathbb{R}$ is said to be L-Lipschitz smooth for some positive real constant $L > 0$ if and only if

Fig. 2.1 Illustration of Lipschitz continuity. If a scalar function $F(x)$ is K-Lipschitz continuous, then for any point x, the function lies inside the region bounded by lines of slope K and $-K$ that pass through the point $(x, F(x))$

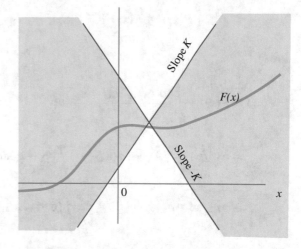

Fig. 2.2 Illustration of Lipschitz smoothness. If a scalar function $F(x)$ is L-Lipschitz smooth, then for any point x, the function lies inside the shaded region shown in the picture, as specified by (2.2)

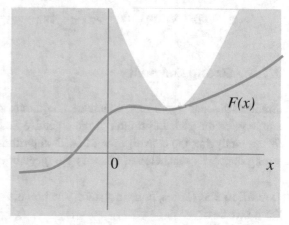

$$|\nabla F(\mathbf{x}_1) - \nabla F(\mathbf{x}_2)| \leq L|\mathbf{x}_1 - \mathbf{x}_2| \quad \text{for all } \mathbf{x}_1, \mathbf{x}_2 \in \mathbb{R}^d \tag{2.2}$$

that is, the gradient $\nabla F(\mathbf{x})$ is L-Lipschitz-continuous.

The intuition behind Lipschitz smoothness is that the rate of change of the function F is bounded by L. In most SGD convergence analyses, we assume that the objective function F is Lipschitz smooth to ensure that it does not change arbitrarily quickly. In particular, the SGD convergence analyses presented in this book will use the following inequality which is a consequence of Lipschitz smoothness (Fig. 2.2).

Lemma 2.1 *If a function is L-Lipschitz smooth, then for any $\mathbf{x}_1, \mathbf{x}_2 \in \mathbb{R}^d$ it satisfies the following upper bound:*

$$F(\mathbf{x}_1) \leq F(\mathbf{x}_2) + \nabla F(\mathbf{x}_2)^\top (\mathbf{x}_1 - \mathbf{x}_2) + \frac{L}{2} \|\mathbf{x}_1 - \mathbf{x}_2\|_2^2 \qquad (2.3)$$

Proof

$$F(\mathbf{x}_1) = F(\mathbf{x}_2) + \int_0^1 \frac{\partial F(\mathbf{x}_2 + t(\mathbf{x}_1 - \mathbf{x}_2))}{\partial t} dt \qquad (2.4)$$

$$= F(\mathbf{x}_2) + \int_0^1 \nabla F(\mathbf{x}_2 + t(\mathbf{x}_2 - \mathbf{x}_1))^\top (\mathbf{x}_1 - \mathbf{x}_2) dt \qquad (2.5)$$

$$= F(\mathbf{x}_2) + \nabla F(\mathbf{x}_2)^\top (\mathbf{x}_2 - \mathbf{x}_1) + \int_0^1 [\nabla F(\mathbf{x}_2 + t(\mathbf{x}_1 - \mathbf{x}_2)) - \nabla F(\mathbf{x}_2)^\top]^\top (\mathbf{x}_1 - \mathbf{x}_2) dt$$

$$(2.6)$$

$$\leq F(\mathbf{x}_2) + \nabla F(\mathbf{x}_2)^\top (\mathbf{x}_1 - \mathbf{x}_2) + \frac{L}{2} \|\mathbf{x}_1 - \mathbf{x}_2\|_2^2 \qquad (2.7)$$

2.1.3   Strong Convexity

Gradient descent algorithms, which are ubiquitous in machine learning, are guaranteed to converge to the global optimum of the objective function $F(\mathbf{x})$ only for convex functions. When analyzing the convergence of SGD algorithms, we will also use a property called strong convexity that helps us quantify the speed of convergence.

Definition 2.5 (*Strong convexity*) If a function is c-strongly convex, then for any $\mathbf{x}_1, \mathbf{x}_2 \in \mathbb{R}^d$ it satisfies:

$$F(\mathbf{x}_1) \geq F(\mathbf{x}_2) + \nabla F(\mathbf{x}_2)^\top (\mathbf{x}_1 - \mathbf{x}_2) + \frac{c}{2} \|\mathbf{x}_1 - \mathbf{x}_2\|_2^2 \qquad (2.8)$$

Every convex function is 0-strongly convex, a special case of the class of strongly convex functions defined above.

Figure 2.3 illustrates the difference between convexity and strong convexity for a one-dimensional function $F(x)$. A convex function always lies above a tangent drawn at any point $(x, F(x))$. On the other hand, if $F(x)$ is strongly convex, then it has to lie above a quadratic function drawn at that point.

A consequence of strong convexity is the Polyak-Lojasiewicz (PL) inequality stated and proved below.

Lemma 2.2 (Polyak-Lojasiewicz (PL) inequality) *If a function is c-strongly convex, then for any $\mathbf{x} \in \mathbb{R}^d$ it satisfies the following lower bound:*

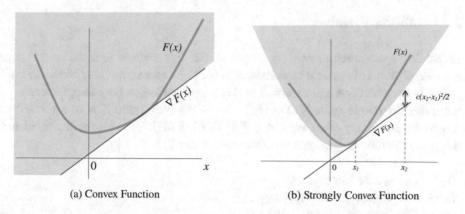

(a) Convex Function   (b) Strongly Convex Function

Fig. 2.3 Illustration of a convex function and a c-strongly convex function

$$2c(F(\mathbf{x}) - F(\mathbf{x}^*)) \leq \|\nabla F(\mathbf{x})\|_2^2 \tag{2.9}$$

Proof The above result follows from Definition 2.5 by minimizing both sides of (2.8) with respect to $\mathbf{x}_1$. The left hand side is minimized by setting $\mathbf{x}_1 = \mathbf{x}^*$. To minimize the right hand side, we take the gradient with respect to $\mathbf{x}_1$ and set it to zero.

$$\nabla_{\mathbf{x}_1} \left[F(\mathbf{x}_2) + \nabla F(\mathbf{x}_2)^\top (\mathbf{x}_1 - \mathbf{x}_2) + \frac{c}{2} \|\mathbf{x}_1 - \mathbf{x}_2\|_2^2 \right] = 0 \tag{2.10}$$

$$[\nabla F(\mathbf{x}_2) + c(\mathbf{x}_1 - \mathbf{x}_2)] = 0 \tag{2.11}$$

$$\mathbf{x}_1 = \mathbf{x}_2 - \frac{1}{c} \nabla F(\mathbf{x}_2) \tag{2.12}$$

Substituting this value of $\mathbf{x}_1$ in right hand side of (2.8) we get:

$$F(\mathbf{x}^*) \geq F(\mathbf{x}_2) - \frac{1}{c} \|\nabla F(\mathbf{x}_2)\|_2^2 + \frac{c}{2} \left\| -\frac{1}{c} \nabla F(\mathbf{x}_2) \right\|_2^2 \tag{2.13}$$

The result follows from simplifying and rearranging the terms in the above equation.

2.2   Probability Review

Next, let us review from probability concepts. We will use these in analyzing the runtime per iteration of the various distributed SGD algorithms covered in this book.

2.2.1 Random Variable

A probability space is represented by a tuple of $(\Omega, \mathcal{F}, \mathcal{P})$ where Ω is the sample space consisting of the set of possible outcomes, $\mathcal{F}$ is the event space consisting of all subsets of Ω, and $\mathcal{P}$ is a probability measure that maps events to probabilities. For example, for one roll of a fair die, the sample space is $\Omega = \{1, 2, \ldots, 6\}$, the event space $\mathcal{F}$ is all the 2^6 subsets of Ω and the probability of an event $A \in \mathcal{F}$ is $\mathcal{P}(A) = |A|/6$. The probability measure $\mathcal{P}$ must satisfy the following axioms for any events A and B:

1. $0 \leq \Pr(A) \leq 1$ for any event A
2. $\Pr(\emptyset) = 0$, where $\emptyset$ is the empty set
3. $\Pr(A \cup B) = \Pr(A) + \Pr(B) - \Pr(A \cap B)$

A real-valued random variable X is a function that maps Ω to the real line $\mathbb{R}$. If it takes a set of discrete values $\{x_1, x_2, \ldots, x_R\} \in \mathbb{R}$, then it is referred to as a discrete random variable and if X takes a continuous set of values $\mathcal{S} \in \mathcal{R}$ then it is referred to as a continuous random variable. Each random variable is associated with a cumulative distribution function (CDF) denoted by $F_X(x)$. For each $x \in \mathbb{R}$, $F_X(x) = \Pr(X \leq x)$, the probability that X is less than or equal to x. By definition $F_X(-\infty) = 0$ and $F_X(\infty) = 1$. A related function is the tail distribution function or the complementary cumulative distribution function (CCDF) denoted by $\bar{F}_X(x) = 1 - F_X(x) = \Pr(X > x)$.

To determine the probability that X takes a specific value x, we define its probability density function (PDF) $f_X(x) = F_X'(x)$, the derivative of the cumulative distribution function $F_X(x)$. For any set $\mathcal{S} \in \mathbb{R}$, we have $\Pr(X \in \mathcal{S}) = \int_{x \in \mathcal{A}} f_X(x)$, and $\int_{-\infty}^{\infty} f_X(x)dx = 1$. If X is discrete, then instead of the probability density function we use the probability mass function (PMF), where $\Pr(X = x_i) = p_i$ is the probability that X takes the value x_i and $\sum_{i=1}^{R} \Pr(X = x_i) = 1$.

2.2.2 Expectation and Variance

Instead of specifying the entire probability distribution of a random variable X, it is often convenient to characterize it in terms of its average value and the deviation around the average, which are captured by the expectation and variable respectively. The expected value or the mean of a random variable are defined as

$$\mathbb{E}[X] = \int_{x \in \mathbb{R}} x f_X(x)dx \quad \text{for continuous } X \tag{2.14}$$

$$= \sum_x x \Pr(X = x) \quad \text{for discrete } X \tag{2.15}$$

For a non-negative random variables $X \geq 0$, it is often convenient to compute the expectation in terms of the tail distribution function $\bar{F}_X(x) = \Pr(X > x)$:

$$\mathbb{E}[X] = \int_{x \in \mathbb{R}} Pr(X > x)dx \quad \text{for non-negative } X \tag{2.16}$$

The deviation of a random variable around its expected value is captured by the variance Var $[X]$ which is defined as:

$$\text{Var}[X] = \int_{x \in \mathbb{R}} (x - \mathbb{E}[X])^2 f_X(x)dx \tag{2.17}$$

$$= \mathbb{E}[X^2] - (\mathbb{E}[X])^2 \tag{2.18}$$

2.2.3 Some Canonical Random Variables

Here are some common random variables that will use in the book:

1. **Bernoulli** $\mathcal{B}(p)$: The sample space $\Omega = \{0, 1\}$, representing two possible outcomes. A coin toss is the canonical example of the Bernoulli random variable. If X is Bernoulli with bias p, then

$$X = \begin{cases} 1 & \text{with probability } p \\ 0 & \text{otherwise} \end{cases} \tag{2.19}$$

   The mean and variance of the Bernoulli distribution are $\mathbb{E}[X] = p$ and Var $[X] = p(1 - p)$ respectively.

2. **Geometric** Geom(p): The geometric random variable represents the first occurrence of success (or failure) in a sequence of independent Bernoulli trials. If the probability of success of each trial is p, then a geometric random variable X has the probability mass function

$$\Pr(X = k) = (1 - p)^{k-1}p \quad \text{for } k = 1, 2, \ldots, \infty. \tag{2.20}$$

   For the coin toss example, the number of tosses until the first heads occurs is geometrically distributed. The mean and variance of the Geometric random variable are

$$\mathbb{E}[X] = \frac{1}{p} \quad \text{and Var}[X] = \frac{1 - p}{p} \tag{2.21}$$

   respectively.

3. **Exponential** Exp(λ): The exponential random variable is a non-negative random variable, which is often used to model delays such as the time taken to execute a job in

computer systems. We will use it heavily in this book when analyzing the runtime of distributed SGD algorithms. The probability density function (PDF) of the exponential random variable for rate $\lambda > 0$ is

$$f_X(x) = \lambda e^{-\lambda x} \text{ for } x \geq 0 \tag{2.22}$$

The mean and variance of the exponential random variable are $\mathbb{E}[X] = \frac{1}{\lambda}$ and $\text{Var}[X] = \frac{1}{\lambda^2}$ respectively.

4. **Gaussian** $\mathcal{N}(\mu, \sigma^2)$: The Gaussian random variable is used to model noise in any random phenomenon, and its probability density function (PDF) is given by:

$$f_X(x) = \frac{1}{\sqrt{2\pi\sigma^2}} e^{-\frac{(x-\mu)^2}{2\sigma^2}} \tag{2.23}$$

The mean and variance of the exponential random variable are $\mathbb{E}[X] = \mu$ and $\text{Var}[X] = \frac{1}{\mu^2}$ respectively. One of the reasons for its ubiquity is the Central Limit Theorem, which shows that the average of a large number of realization of a random variables with any distribution with finite mean and finite variance converges to the Gaussian distribution.

5. **Pareto** $\text{Pareto}(x_m, \alpha)$: The Pareto distribution is a power-law distribution that is used to model many observable phenomenon such as cloud server delays, error or failure rates of hardware components, wealth distribution of a population, online file size, etc. Its probability density function (PDF) is given by:

$$f_X(x) = \begin{cases} \frac{\alpha x_m^\alpha}{x^{\alpha+1}} & x \geq x_m \\ 0 & x < x_m \end{cases} \tag{2.24}$$

The parameters x_m and α are referred to as scale and shape parameters respectively. Larger α implies a faster decay of the tail. The mean and variance of a Pareto random variable are given by:

$$\mathbb{E}[X] = \begin{cases} \infty & \alpha \leq 1 \\ \frac{\alpha x_m}{\alpha-1} & \alpha > 1 \end{cases} \tag{2.25}$$

$$\text{Var}[X] = \begin{cases} \infty & \alpha \leq 2 \\ \frac{\alpha x_m^2}{(\alpha-1)^2(\alpha-2)} & \alpha > 2 \end{cases} \tag{2.26}$$

2.2.4 Bayes Rule and Conditional Probability

For two events A and B, the conditional probability A given that event B has occurred is denoted as $\Pr(A|B)$. Bayes theorem connects that conditional probabilities $\Pr(A|B)$ and $\Pr(B|A)$ as follows:

$$\Pr(A|B) = \frac{\Pr(A \cap B)}{\Pr(B)} = \frac{\Pr(B|A)\Pr(A)}{\Pr(B)} \tag{2.27}$$

A consequence of Bayes theorem is the concept of the residual time of a random variable, which we formally define as follows.

Definition 2.6 (*Residual time of a random variable*) For a non-negative random variable $X \geq 0$ that denotes the time taken to complete a task, the residual time Y is the time remaining given that t units of time have elapsed. The random variable $Y = (X - t)|(X > t)$ and its probability distribution is:

$$F_Y(y) = \frac{F_X(t + y)}{\Pr(X > t)} \tag{2.28}$$

Depending on how the tail distribution $\Pr(Y > y)$ of the residual time $Y = (X - t)|(X > t)$ compares with the tail distribution $\Pr(X > x)$ of X itself, we can define two special classes of random variables called new-longer-than-used and new-shorter-than-used.

Definition 2.7 (*New-longer-than-used and new-shorter-than-used*) A random variable X is said to have a new-longer-than-used distribution if the tail distribution of the following holds for all $t, x \geq 0$:

$$\Pr(X > x + t|X > t) \leq \Pr(X > x). \tag{2.29}$$

On the other hand, X is said to have a new-shorter-than-used distribution if the tail distribution of the following holds for all $t, x \geq 0$:

$$\Pr(X > x + t|X > t) \geq \Pr(X > x). \tag{2.30}$$

To understand the intuition behind this notion, let a random variable X denote the computational time taken to perform a task. Suppose that the task has been running for t units of time but has not finished, and the scheduler needs to decide whether to keep the task running or abort it and launch a new copy. If X has a new-longer-than-used distribution, then a new copy is expected to take longer than waiting for the already running task to finish. Most of the continuous distributions we encounter like normal, shifted-exponential, gamma, beta are new-longer-than-used. On the other hand, the hyper-exponential distribution (mixture of exponentials) is new-shorter-than-used. If the computation time X of a task is new-shorter-than-used, then launching a new copy is better than keeping an old copy of the task running.

The exponential random variable is a special random variable—it is the only continuous random variable that is both new-longer-than-used and new-shorter-than-used. For the exponential distribution, (2.29) and (2.30) holds with equality. Thus, the exponential distribution has the *memoryless* property, which means that the residual time $Y = (X - t)|(X > t)$ is also an exponential random variable, as shown below.

$$F_Y(y) = \frac{e^{-\lambda(y+t)}}{e^{-\lambda t}} = e^{-\lambda y} \quad \text{for all } y, t \geq 0 \tag{2.31}$$

This implies that a fresh realization of the exponential random variable X, and a residual version with time t elapsed are statistically identical. Among discrete random variables, the geometric random variable is the analogue of the exponential and it is the only discrete random variable to have the memoryless property.

2.3   Order Statistics

Consider n independent and identically distributed (i.i.d.) realizations $X_1, \ldots, X_n$ of a random variable X, then the order statistics $X_{1:n}, X_{2:n}, \ldots, X_{n:n}$ denote the realizations sorted in increasing order, that is $X_{1:n} = \min(X_1, \ldots, X_n)$ and $X_{n:n} = \max(X_1, \ldots, X_n)$. The order statistics $X_{k:n}$ themselves are random variables and we can express their probability distribution and probability density functions in terms of the distribution $F_X(x)$ of X:

$$F_{X_{k:n}}(x) = \sum_{j=k}^{n} \binom{n}{j} [F_X(x)]^j [1 - F_X(x)]^{n-j} \tag{2.32}$$

$$f_{X_{k:n}}(x) = \frac{n!}{(k-1)!(n-k)!} f_X(x) [F_X(x)]^{k-1} [1 - F_X(x)]^{n-k} \tag{2.33}$$

The two extreme special cases, the maximum order statistic $X_{n:n}$ and the minimum order statistic $X_{1:n}$ have cumulative distribution functions:

$$F_{X_{n:n}}(x) = \Pr(\max(X_1, \ldots, X_n) \leq x) = (F_X(x))^n \tag{2.34}$$

$$F_{X_{1:n}}(x) = \Pr(\min(X_1, \ldots, X_n) \leq x) = 1 - (1 - F_X(x))^n. \tag{2.35}$$

2.3.1   Order Statistics of the Exponential Distribution

Let us determine the distribution of the k-th order statistic $X_{k:n}$ of the exponential random variable $X \sim \text{Exp}(\lambda)$. Firstly observe that $X_{1:n}$, the minimum of n exponentials is an exponential with rate $n\lambda$, as given by its derived probability distribution below:

$$\Pr(X_{1:n} > x) = \Pr(\min(X_1, X_2, \ldots, X_n > x) \tag{2.36}$$

$$= \Pr(X_i > x)^n \tag{2.37}$$

$$= e^{-n\lambda} \tag{2.38}$$

Fig. 2.4 Illustration of order
Statistics of exponential
random variables

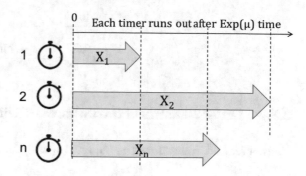

Due to the memoryless property of the exponential, after $X_{1:n}$ time elapses the residual time of the remaining $n-1$ exponentials is also exponentially distributed. Therefore, after $X_{2:n}$ is $X_{1:n}$ plus the minimum of $n-1$ exponential random variables, as illustrated in Fig. 2.4. Continuing to use the memoryless property, we can show that the k-th order statistic

$$X_{k:n} = \sum_{i=1}^{k} \frac{Z_i}{n - i + 1} \tag{2.39}$$

where Z_i's are i.i.d. exponential random variables with rate λ. The expected value of $X_{k:n}$ is given by

$$\mathbb{E}[X_{k:n}] = \sum_{i=1}^{k} \frac{1}{\lambda(n - i + 1)} \tag{2.40}$$

$$= \frac{H_n - H_{n-k}}{\lambda} \tag{2.41}$$

where $H_n = \sum_{i=1}^{n} 1/i$, the n-th Harmonic number. For large n, $H_n \approx \log n$.

By using the fact the the variance of a sum of independent random variables is a sum of their variances, the variance of $X_{k:n}$ is given by:

$$\text{Var}[X_{k:n}] = \sum_{i=1}^{k} \sum_{i=1}^{k} \text{Var}\left[\frac{Z_i}{n - i + 1}\right] \tag{2.42}$$

$$= \sum_{i=1}^{k} \frac{1}{\lambda^2 (n - i + 1)^2} \tag{2.43}$$

$$= \frac{1}{\lambda^2}\left(H_n^{(2)} - H_{n-k}^{(2)}\right) \tag{2.44}$$

where $H_n^{(2)} = \sum_{i=1}^{n} 1/i^2$, the generalized Harmonic number.

For the special case of the maximum order statistic $X_{n:n}$, the mean and variance are given by:

$$\mathbb{E}[X_{n:n}] = \frac{H_n}{\lambda} \approx \frac{\log n}{\lambda} \tag{2.45}$$

$$\text{Var}[X_{n:n}] = \frac{1}{\lambda^2}\left(H_n^{(2)}\right) = \frac{n(n+1)(2n+1)}{6\lambda^2} = O(n^3). \tag{2.46}$$

2.3.2 Order Statistics of the Uniform Distribution

Consider n i.i.d. realizations $U_1, U_2, \ldots, U_n$ of the uniform distribution

$$f_U(u) = \begin{cases} 1 & \text{for } u \in [0, 1] \\ 0 & \text{otherwise} \end{cases}. \tag{2.47}$$

The k-th order statistic $U_{k:n}$ follows the Beta distribution $\text{Beta}(k, n+1-k)$ and its probability distribution function (pdf) is given by:

$$f_{U_{k:n}}(u) = \frac{n!}{(k-1)!(n-k)!}u^{k-1}(1-u)^{n-k} \tag{2.48}$$

The mean of the k-th order statistics is given by $\mathbb{E}[U_{k:n}] = \frac{k}{n+1}$.

2.3.3 Asymptotic Distribution of Quantiles

For a large number of samples n from a distribution F_X, it is interesting to understand the asymptotic distribution of the quantiles, that is the order statistic $X_{\lceil np \rceil:n}$ for $p \in (0, 1)$. It can be shown that as $n \to \infty$, $X_{\lceil np \rceil:n}$ converges in distribution to a Gaussian (Normal) distribution:

$$X_{\lceil np \rceil:n} \sim \mathcal{N}\left(F^{-1}(p), \frac{p(1-p)}{n[f(F^{-1}(p))]^2}\right). \tag{2.49}$$

Summary

In this chapter we reviewed calculus concepts, in particular, strong convexity and Lipschitz smoothness, which will be used in the convergence analysis of SGD and its variants in the upcoming chapters of the book. We also reviewed the basics of probability theory and discussed the concept of order statistics of random variables. Order statistics will be used to analyze the runtime per iteration of various distributed SGD algorithms that we will study in this book.

Problems

1. Identify whether the following functions are Lipschitz continuous, and if they are, then determine the smallest corresponding Lipschitz continuity parameter K: (1) $F(x) = \sin(x)$, and (2) $F(x) = x^2$.
2. Identify whether the following functions are Lipschitz smooth, and if they are, determine the smallest corresponding Lipschitz smoothness parameter L: 1) $F(x) = \frac{1}{2}\|\mathbf{Ax} - \mathbf{y}\|^2$, and 2) $F(x) = x^2$
3. Consider the Pareto distribution, whose cumulative distribution function is given by:

$$F_X(x) = \begin{cases} 1 - \left(\frac{x_m}{x}\right)^\alpha & x \geq x_m \\ 0 & x < x_m \end{cases} \tag{2.50}$$

   Compare the mean $\mathbb{E}[X]$ with the mean of the residual time $\mathbb{E}[(X - t)|X > t]$ for some elapsed time $t \geq 0$. Is a new realization of the Pareto random variable expected to be longer than the expected residual time of an old realization?
4. Let X be a random variable with mean μ and variance σ^2 (assuming $\sigma > 0$). Let ϵ be a random variable drawn from standard Gaussian $\mathcal{N}(0, 1)$ such that X and ϵ are independent. What is the variance of the random variable $Y = X/\sigma + \epsilon$?
5. Find the probability distribution, expectation and variance of the k-th order statistic $X_{k:n}$ of n i.i.d. realizations of a shifted exponential distributed $X \sim \Delta + \mathsf{Exp}(\lambda)$ for some $\Delta > 0$.
6. Find the probability distribution, expectation and variance of the minimum order statistic $X_{1:n}$ of n i.i.d. realizations of the Pareto distribution $X \sim \mathsf{Pareto}(x_m, \alpha)$.

Convergence of SGD and Variance-Reduced Variants

In this chapter we will analyze the convergence of gradient descent (GD) and stochastic gradient descent (SGD) to determine the number of iterations required to reach a target error. Because it uses noisy estimates of the true gradient, SGD has slower convergence than GD. Therefore, recent works have proposed variance-reduced versions of SGD that improve its convergence rate. We will review some of these variance-reduced variants of SGD.

3.1 Gradient Descent (GD) Convergence

Recall from Chap. 1 that GD descent is an iterative algorithm that minimize the loss function $F(\mathbf{w})$ with respect to the parameter vector $\mathbf{w}$. It starts with a randomly initialized $\mathbf{w}_0$, and uses the following update rule:

$$\mathbf{w}_{t+1} = \mathbf{w}_t - \eta \nabla F(\mathbf{w}_t) \tag{3.1}$$

where η is the step size or the learning rate. The convergence speed of GD depends on the choice of η and the properties of the loss function $F(\mathbf{w})$. As with most other gradient-based methods, for non-convex objectives, GD may get stuck in a local minima. However, for strongly convex and smooth $F(\mathbf{w})$ and small enough η, it is guaranteed to convergence to the optimal x^*.

Theorem 3.1 (Convergence of gradient descent (GD)) *For a c-strongly convex and L-smooth function, if the learning rate $\eta < \frac{1}{L}$ and the starting point is $\mathbf{w}_0$, then $F(\mathbf{w}_t)$ after t gradient descent iterations is bounded as*

$$F(\mathbf{w}_t) - F(\mathbf{w}^*) \leq (1 - \eta c)^t (F(\mathbf{w}_0) - F(\mathbf{w}^*)) \tag{3.2}$$

© The Author(s), under exclusive license to Springer Nature Switzerland AG 2023
G. Joshi, *Optimization Algorithms for Distributed Machine Learning*,
Synthesis Lectures on Learning, Networks, and Algorithms,
https://doi.org/10.1007/978-3-031-19067-4_3

Proof Recall that if a function is L-Lipschitz smooth, then for any $\mathbf{x}_1, \mathbf{x}_2 \in \mathbb{R}^d$ it satisfies:

$$F(\mathbf{x}_1) \leq F(\mathbf{x}_2) + \nabla F(\mathbf{x}_2)^\top (\mathbf{x}_1 - \mathbf{x}_2) + \frac{L}{2} \|\mathbf{x}_1 - \mathbf{x}_2\|_2^2 \tag{3.3}$$

Replacing $\mathbf{x}_1$ with $\mathbf{w}_{t+1}$ and $\mathbf{x}_2$ with $\mathbf{w}_t$ respectively we have

$$F(\mathbf{w}_{t+1}) - F(\mathbf{w}_t) \leq \nabla F(\mathbf{w}_t)^\top (\mathbf{w}_{t+1} - \mathbf{w}_t) + \frac{L}{2} \|\mathbf{w}_{t+1} - \mathbf{w}_t\|^2 \tag{3.4}$$

$$\leq \nabla F(\mathbf{w}_t)^\top (-\eta \nabla F(\mathbf{w}_t)) + \frac{L}{2} \| - \eta \nabla F(\mathbf{w}_t)\|^2 \tag{3.5}$$

$$\leq \eta \left(1 - \frac{L}{2}\eta\right) (-\|\nabla F(\mathbf{w}_t)\|^2) \tag{3.6}$$

Now using the Polyak-Lojasiewicz (PL) inequality $2c(F(\mathbf{w}) - F(\mathbf{w}^*)) \leq \|\nabla F(\mathbf{w})\|^2$, which is a consequence of c-strong-convexity of $F(\mathbf{w})$ we have

$$F(\mathbf{w}_{t+1}) - F(\mathbf{w}_t) \leq \eta \left(1 - \frac{L}{2}\eta\right) (-2c(F(\mathbf{w}_t) - F(\mathbf{w}^*))) \tag{3.7}$$

Assume that $\eta < \frac{1}{L}$. Then $\left(1 - \frac{L}{2}\eta\right) \geq \frac{1}{2}$. Thus,

$$F(\mathbf{w}_{t+1}) - F(\mathbf{w}_t) \leq -\eta c(F(\mathbf{w}_t) - F(\mathbf{w}^*)) \tag{3.8}$$

$$F(\mathbf{w}_{t+1}) - F(\mathbf{w}^*) + F(\mathbf{w}^*) - F(\mathbf{w}_t) \leq -\eta c(F(\mathbf{w}_t) - F(\mathbf{w}^*)) \tag{3.9}$$

$$F(\mathbf{w}_{t+1}) - F(\mathbf{w}^*) \leq (1 - \eta c)(F(\mathbf{w}_t) - F(\mathbf{w}^*)) \tag{3.10}$$

Continuing recursively, we will have

$$F(\mathbf{w}_{t+1}) - F(\mathbf{w}^*) \leq (1 - \eta c)^2 (F(\mathbf{w}_{t-1}) - F(\mathbf{w}^*)) \tag{3.11}$$

$$\vdots \tag{3.12}$$

$$\leq (1 - \eta c)^{t+1} (F(\mathbf{w}_0) - F(\mathbf{w}^*)) \tag{3.13}$$

3.1.1   Effect of Learning Rate and Other Parameters

The above analysis shows that as $t \to \infty$, $F(\mathbf{w}_t)$ converges to the optimal value $F(\mathbf{w}^*)$ and the optimality gap shrinks exponentially fast due to the multiplicative factor $(1 - \eta c) < 1$. The rate of convergence of $F(\mathbf{w})$ to its optimal value $F(\mathbf{w}^*)$ increases with η and c, because it will result in a smaller value of the multiplicative factor $(1 - \eta c)$. Figure 3.1 illustrates the effect of η on the convergence of batch gradient descent for a simple linear regression problem. In general, η is a hyperparameter that requires manual tuning in order to set it to a value that is best suited to the dataset and the objective function in question. A smaller value of the Lipschitz constant L also results in faster convergence because η can be increased up to

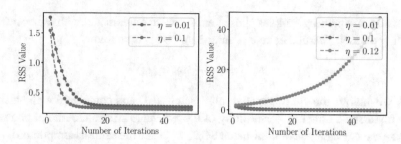

Fig. 3.1 Illustration of batch gradient descent convergence for a linear regression problem, where the y-axis is the residual sum of squares (RSS) error. When the learning rate η increases from $\eta = 0.01$ to $\eta = 0.1$ we get faster convergence but increasing η further to $\eta = 0.12$ causes the error to diverge

$1/L$. Unlike the learning rate η, the Lipschitz constant L and the strong convexity parameter c are properties of the objective function and cannot be controlled by the optimization algorithm.

3.1.2 Iteration Complexity

More concretely, let us determine how many iterations do it takes to reach error $F(\mathbf{w}_t) - F(\mathbf{w}^*) = \epsilon$. We can evaluate it as follows.

$$(1 - \eta c)^t (F(\mathbf{w}_0) - F(\mathbf{w}^*)) \le \epsilon \tag{3.14}$$

$$t \log(1 - \eta c) + \log(F(\mathbf{w}_0) - F(\mathbf{w}^*)) \le \log(\epsilon) \tag{3.15}$$

$$t \log(1/(1 - \eta c)) - \log(F(\mathbf{w}_0) - F(\mathbf{w}^*) \ge \log(\frac{1}{\epsilon}) \tag{3.16}$$

$$t = O\left(\log(\frac{1}{\epsilon})\right) \tag{3.17}$$

This convergence rate is often called *linear convergence*. It is considered to be a fast convergence speed because of the log term. For instance, to achieve an error $\epsilon = 10^{-m}$, we only need $O(m)$ iterations.

3.2 Convergence Analysis of Mini-batch SGD

Since many practical ML problems use the empirical risk function $F(\mathbf{w}) = \frac{1}{N} \sum_{n=1}^{N} \ell(\mathbf{w}, \xi_n)$, it is too expensive to compute the full gradient $\nabla F(\mathbf{w})$ of the objective function because it requires processing each of the N samples of a large training dataset in every iteration. Mini-batch stochastic gradient descent (SGD) is a computationally efficient alternative where the gradient $\nabla F(\mathbf{w}_t)$ is replaced by a noisy estimate $g(\mathbf{w}_t, \xi_t) = \sum_{l=1}^{b} \nabla \ell(\mathbf{w}_t, \xi_{t,l})/b$ computed using a batch of b training samples chosen uniformly at random and with replacement from the training dataset. The update rule of mini-batch SGD is as follows:

$$\mathbf{w}_{t+1} = \mathbf{w}_t - \eta g(\mathbf{w}_t, \xi_t) \tag{3.18}$$

Next, we analyze the convergence of mini-batch SGD and compare it with batch GD in order to understand how the stochasticity of the gradients affects the rate of convergence. To analyze the convergence of mini-batch SGD, we make the same assumptions of c-strong convexity and L-Lipschitz smoothness. In addition, we also need assumptions on the mini-batch stochastic gradients $g(\mathbf{w}; \xi)$:

- **Unbiased Estimate**: The mini-batch stochastic gradient $g(\mathbf{w}; \xi)$ is an unbiased estimate of $\nabla F(\mathbf{w})$, that is,

$$\mathbb{E}_\xi[g(\mathbf{w}; \xi)] = \nabla F(\mathbf{w})$$

- **Bounded Variance**: The gradient $\nabla \ell(\mathbf{w}, \xi_l)$ of the l-th sample in the mini-batch has bounded variance, that is,

$$\text{Var}(\nabla \ell(\mathbf{w}; \xi)) \leq \sigma^2 \tag{3.19}$$

This implies the following bounds on the variance of the mini-batch stochastic gradient $g(\mathbf{w}; \xi)$:

$$\mathbb{E}_\xi[\|g(\mathbf{w}; \xi)\|^2] - \|\mathbb{E}_\xi[g(\mathbf{w}; \xi)]\|^2 \leq \frac{\sigma^2}{b} \tag{3.20}$$

$$\mathbb{E}_\xi[\|g(\mathbf{w}; \xi)\|^2] \leq \|\mathbb{E}_\xi[g(\mathbf{w}; \xi)]\|^2 + \frac{\sigma^2}{b} \tag{3.21}$$

$$\mathbb{E}_\xi[\|g(\mathbf{w}; \xi)\|^2] \leq \|\nabla F(\mathbf{w})\|^2 + \frac{\sigma^2}{b} \tag{3.22}$$

Under these assumptions, we have the following result on the convergence of stochastic gradient descent.

Theorem 3.2 (Convergence of mini-batch SGD) *For a c-strongly convex and L-smooth function satisfying the unbiased gradient estimate and bounded variance assumptions listed above, if the learning rate $\eta < \frac{1}{L}$ and the starting point is $\mathbf{w}_0$, then $\mathbb{E}[F(\mathbf{w}_t)]$ after t mini-batch SGD iterations is bounded as*

$$\mathbb{E}[F(\mathbf{w}_t)] - F(\mathbf{w}^*) - \frac{\eta L \sigma^2}{2cb} \leq (1 - \eta c)^t \left(\mathbb{E}[F(\mathbf{w}_0)] - F(\mathbf{w}^*) - \frac{\eta L \sigma^2}{2cb} \right) \tag{3.23}$$

Proof Starting with the Lipschitz smoothness of the objective function we have

$$F(\mathbf{w}_{t+1}) - F(\mathbf{w}_t) \leq \nabla F(\mathbf{w}_t)^\top (\mathbf{w}_{t+1} - \mathbf{w}_t) + \frac{L}{2} \|\mathbf{w}_{t+1} - \mathbf{w}_t\|^2 \tag{3.24}$$

$$\leq \nabla F(\mathbf{w}_t)^\top (-\eta g(\mathbf{w}_t, \xi_t)) + \frac{L}{2} \| - \eta g(\mathbf{w}_t, \xi_t)\|^2 \tag{3.25}$$

Taking expectation on both sides with respect to the stochasticity ξ_t in the t-th iteration,

$$\mathbb{E}\left[F(\mathbf{w}_{t+1})\right] - F(\mathbf{w}_t) \leq -\eta \nabla F(\mathbf{w}_t)^\top \mathbb{E}\left[g(\mathbf{w}_t, \xi_t)\right] + \frac{\eta^2 L}{2} \mathbb{E}\left[\|g(\mathbf{w}_t, \xi_t)\|^2\right] \tag{3.26}$$

$$\leq -\eta \nabla F(\mathbf{w}_t)^\top \mathbb{E}\left[g(\mathbf{w}_t, \xi_t)\right] + \frac{\eta^2 L}{2} \mathbb{E}\left[\|g(\mathbf{w}_t, \xi_t)\|^2\right] \tag{3.27}$$

Using the unbiased gradient assumption $\mathbb{E}_\xi[g(\mathbf{w}; \xi)] = \nabla F(\mathbf{w})$ and the bounded variance assumption $\mathbb{E}_\xi[\|g(\mathbf{w}; \xi)\|^2] \leq \|\nabla F(\mathbf{w})\|^2 + \sigma^2$ we have

$$\mathbb{E}\left[F(\mathbf{w}_{t+1})\right] - F(\mathbf{w}_t) \leq -\eta \|\nabla F(\mathbf{w}_t)\|^2 + \frac{\eta^2 L}{2}(\|\nabla F(\mathbf{w})\|^2 + \frac{\sigma^2}{b}) \tag{3.28}$$

$$\leq \eta \left(1 - \frac{L}{2}\eta\right)(-\|\nabla F(\mathbf{w}_t)\|^2) + \frac{\eta^2 \sigma^2 L}{2b} \tag{3.29}$$

Now using the strong convexity property $2c(F(\mathbf{w}) - F(\mathbf{w}^*)) \leq \|\nabla F(\mathbf{w})\|^2$ we have

$$\mathbb{E}\left[F(\mathbf{w}_{t+1})\right] - F(\mathbf{w}_t) \leq \eta \left(1 - \frac{L}{2}\eta\right)(2c(F(\mathbf{w}_t) - F(\mathbf{w}^*))) + \frac{\eta^2 \sigma^2 L}{2b} \tag{3.30}$$

$$\leq \eta(-c(\mathbb{E}\left[F(\mathbf{w}_t)\right] - F(\mathbf{w}^*))) + \frac{\eta^2 \sigma^2 L}{2b} \tag{3.31}$$

where in (3.31) we use the assumption that $\eta < \frac{1}{L}$, which implies that $\left(1 - \frac{L}{2}\eta\right) \geq \frac{1}{2}$. Now taking total expectations and subtracting the optimal $F(\mathbf{w}^*)$ from both sides we have

$$\mathbb{E}\left[F(\mathbf{w}_{t+1})\right] - F(\mathbf{w}^*) + F(\mathbf{w}^*) - \mathbb{E}\left[F(\mathbf{w}_t)\right] \leq -\eta c(\mathbb{E}\left[F(\mathbf{w}_t)\right] - F(\mathbf{w}^*)) + \frac{\eta^2 \sigma^2 L}{2b} \tag{3.32}$$

$$\mathbb{E}\left[F(\mathbf{w}_{t+1})\right] - F(\mathbf{w}^*) \leq (1 - \eta c)(\mathbb{E}\left[F(\mathbf{w}_t)\right] - F(\mathbf{w}^*)) + \frac{\eta^2 \sigma^2 L}{2b} \tag{3.33}$$

Subtracting $\frac{\eta L \sigma^2}{2cb}$ from both sides

$$\mathbb{E}\left[F(\mathbf{w}_{t+1})\right] - F(\mathbf{w}^*) - \frac{\eta L \sigma^2}{2c} \leq (1 - \eta c)(\mathbb{E}\left[F(\mathbf{w}_t)\right] - F(\mathbf{w}^*)) + \frac{\eta^2 \sigma^2 L}{2} - \frac{\eta L \sigma^2}{2cb}$$

$$\mathbb{E}\left[F(\mathbf{w}_{t+1})\right] - F(\mathbf{w}^*) - \frac{\eta L \sigma^2}{2cb} \leq (1 - \eta c)\left(\mathbb{E}\left[F(\mathbf{w}_t)\right] - F(\mathbf{w}^*) - \frac{\eta L \sigma^2}{2cb}\right)$$

Applying the above inequality recursively for t iterations starting from $\mathbf{w}_0$, we have the result.   □

3.2.1 Effect of Learning Rate and Mini-batch Size

Similar to GD, the above analysis implies that the rate of convergence of $F(\mathbf{w})$ to its optimal value $F(\mathbf{w}^*)$ increases with η and c, because it will result in a smaller value of the multiplicative factor $(1 - \eta c)$. Figure 3.2 illustrates this phenomenon for the linear regression example. However, unlike GD, as $t \to \infty$, the loss function value $F(\mathbf{w}_t)$ does not converge to the optimal $F(\mathbf{w}^*)$.

The stochasticity of the gradients results in an error floor $\eta L \sigma^2 / 2cb$ which increases with η and the variance bound σ^2. Increasing the mini-batch size b reduces the variance and thus reduces the error floor. In the special case when $b = N$, mini-batch SGD reduces to batch GD and the learning curve will have no stochasticity as we saw in Fig. 3.1.

3.2.2 Iteration Complexity

Due to the non-zero error floor $\eta L \sigma^2 / 2cb$, when we use a constant learning rate η the progress of the SGD algorithm towards the optimal model $\mathbf{w}^*$ stalls after some iterations. In order to achieve zero error floor, the learning rate η is gradually decayed during the training process. With such a decaying learning rate schedule, how many iterations do we need to converge to reach error $F(\mathbf{w}_t) - F(\mathbf{w}^*) = \epsilon$? It can be shown that a learning rate schedule $\eta_t = \frac{\beta}{\gamma + t}$ for some constants $\beta, \gamma > 0$ gives the following convergence result, from which we can infer the iterations complexity.

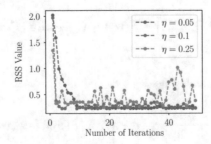

Fig. 3.2 Illustration of stochastic gradient descent convergence for a linear regression problem, where the y-axis is the residual sum of squares (RSS) error. Higher learning rate η yields faster convergence, but also results in a higher error floor

Theorem 3.3 (Convergence of SGD with decaying learning rate) *For a c-strongly convex and L-smooth function, suppose the learning rate $\eta_t = \frac{\beta}{\gamma + t}$ in the t-th iteration for some $\beta > \frac{1}{c}$ and $\gamma > 0$ such that $\eta_1 \leq 1/L$. With the starting point $\mathbf{w}_0$, $F(\mathbf{w}_t)$ after t gradient descent iterations is bounded as*

$$\mathbb{E}\left[F(\mathbf{w}_t) - F(\mathbf{w}^*)\right] \leq \frac{\nu}{\gamma + t} \tag{3.34}$$

where $\nu = \max\left(\frac{\beta^2 L \sigma^2}{2(\beta c - 1)}, (\gamma + 1)(F(\mathbf{w}_0) - F(\mathbf{w}^*))\right)$.

The proof can be found in [1]. Theorem 3.3 implies that the number of iterations required to converge to reach error $F(\mathbf{w}_t) - F(\mathbf{w}^*) = \epsilon$ is $O(\frac{1}{\epsilon})$. For instance, to achieve an error $\epsilon = 10^{-m}$, we only need $O(10^m)$ iterations. This convergence rate is much slower than the $O(\log \frac{1}{\epsilon})$ linear convergence rate achieved by batch gradient descent, as we showed in Sect. 3.1.

3.2.3 Non-convex Objectives

Many practical machine learning problems have non-convex objective functions, which do not satisfy the strongly convex assumption. We can extend the above convergence analysis to non-convex objectives, as we show below. However, this general convergence analysis can only guarantee convergence to a stationary point of the objective function (a local minimum or saddle point) because there is no unique global minimum as in the case of strongly convex objective functions. Instead, we assume that the sequence of function values $F(\mathbf{w}_k)$ is bounded below by a scalar F_{inf}.

Theorem 3.4 (Convergence of mini-batch SGD, non-convex objectives) *For a L-smooth function satisfying the unbiased gradient estimate and bounded variance assumptions listed above, if the learning rate $\eta < \frac{1}{L}$ and the starting point is $\mathbf{w}_1$, then after t mini-batch SGD iterations the expected average of squared gradients of F can be bounded as follows*

$$\mathbb{E}\left[\frac{1}{t}\sum_{k=1}^{t}\|\nabla F(\mathbf{w}_k)\|_2^2\right] \leq \frac{\eta L \sigma^2}{b} + \frac{2(F(\mathbf{w}_0) - F_{inf})}{t\eta} \tag{3.35}$$

Proof We follow the same steps as the proof of Theorem 3.2 starting with the Lipschitz smoothness of the objective function to get

$$\mathbb{E}\left[F(\mathbf{w}_{k+1})\right] - F(\mathbf{w}_k) \leq \eta\left(1 - \frac{L}{2}\eta\right)(-\|\nabla F(\mathbf{w}_k)\|^2) + \frac{\eta^2 \sigma^2 L}{2b} \tag{3.36}$$

$$\leq -\frac{\eta}{2}\|\nabla F(\mathbf{w}_k)\|^2 + \frac{\eta^2 \sigma^2 L}{2b} \tag{3.37}$$

where we use the assumption that $\eta < \frac{1}{L}$, which implies that $\left(1 - \frac{L}{2}\eta\right) \geq \frac{1}{2}$. Summing both sides for $k = 1, \ldots, t$ and dividing by t we get:

$$\frac{\mathbb{E}\left[F(\mathbf{w}_{t+1})\right] - F(\mathbf{w}_1)}{t} \leq -\frac{\eta}{2t} \sum_{k=1}^{t} \|\nabla F(\mathbf{w}_k)\|^2 + \frac{\eta^2 \sigma^2 L}{2b} \tag{3.38}$$

$$\frac{1}{t} \sum_{k=1}^{t} \|\nabla F(\mathbf{w}_k)\|^2 \leq -\frac{2(\mathbb{E}\left[F(\mathbf{w}_{t+1})\right] - F(\mathbf{w}_1))}{\eta t} + \frac{\eta \sigma^2 L}{b} \tag{3.39}$$

$$\frac{1}{t} \sum_{k=1}^{t} \|\nabla F(\mathbf{w}_k)\|^2 \leq -\frac{2(F_{inf} - F(\mathbf{w}_1))}{\eta t} + \frac{\eta \sigma^2 L}{b} \tag{3.40}$$

We can also extend this result to obtain a convergence analysis of SGD with diminishing step-size η for non-convex objective functions. However, we will skip stating that here and refer the readers to [1] for the details.

3.3 Variance-Reduced SGD Variants

In Sect. 3.2 we saw that mini-batch SGD, the computationally efficient version of batch GD that uses noisy stochastic gradients instead of full gradients, has a non-zero error floor due to the gradient noise. In order to achieve convergence to the optimum $F(\mathbf{w}^*)$, we need to decay the learning rate during the course of training. Even with this decay learning rate strategy we need $O(1/\epsilon)$ iterations to reach an ϵ error, which is significantly larger than the $O(\log(1/\epsilon))$ iterations required with batch GD. To bridge this gap, several recent papers have proposed techniques to reduce the variance of the stochastic gradients and achieve the $O(\log(1/\epsilon))$ linear convergence rate. We will discuss three such techniques in this section, namely, Dynamic mini-batch size, Stochastic Average Gradient (SAG) [2] and Stochastic Variance Reduced Gradient (SVRG) [3].

3.3.1 Dynamic Mini-batch Size Schedule

Instead of decaying learning rate, an alternate approach to gradually reducing the gradient noise in order to achieve a zero error floor is to dynamically increase the mini-batch size b during the course of training. In particular, consider the following dynamic mini-batch version of SGD, where

$$\mathbf{w}_{t+1} = \mathbf{w}_t - \eta g(\mathbf{w}_t, \xi_t) \quad \text{where} \tag{3.41}$$

$$g(\mathbf{w}_t, \xi_t) = \frac{1}{b_t} \sum_{j \in \mathcal{S}_t} \nabla \ell(\mathbf{w}_t, \xi_{t,j}) \quad \text{with } b_t = |\mathcal{S}_t| = \lceil \gamma^{t-1} \rceil \tag{3.42}$$

for some $\gamma > 1$. Thus, the mini-batch size b_t is increased at geometrically at the rate γ. The variance of the gradient in the t-th iteration is bounded $\text{Var}(g(\mathbf{w}_t; \xi)) \leq \frac{M}{b_t}$. With this dynamic mini-batch schedule, SGD can achieve a $F(\mathbf{w}_t) - F(\mathbf{w}^*) \leq \epsilon$ error in $O(\log \frac{1}{\epsilon})$, matching the linear convergence rate of batch GD.

While the number of iterations required to reach ϵ error reduces, note that the total number of per-sample gradient computation increases geometrically in each iteration. Over t iterations, the number of gradient computations is proportional to γ^t. Thus, over $O(\log \frac{1}{\epsilon})$ iterations, the number of gradient computations is $O(\frac{1}{\epsilon})$. This is exactly the iterations complexity of SGD, which only performs 1 gradient evaluation per iteration.

3.3.2 Stochastic Average Gradient (SAG)

In each stochastic gradient descent (SGD) chooses a data sample ξ_t uniformly at random from the N training samples and uses the gradient at that sample in lieu of the full gradient computed over the entire dataset. Stochastic Average Gradient (SAG) proposed in [2] maintains previously computed gradients in memory and uses them in place of the $N-1$ missing gradients in each iterations. The SAG update rule is given by:

$$\mathbf{w}_{t+1} = \mathbf{w}_t - \eta \frac{1}{N} \left(\nabla \ell(\mathbf{w}_t, \xi_t) - \mathbf{v}_{t,n} + \sum_{n=1}^{N} \mathbf{v}_{t,n} \right) \tag{3.43}$$

where $\mathbf{v}_n$ is the (potentially outdated) gradient at the n-th sample that is stored in memory. After the t-th iteration, these vectors are updated as follows:

$$\mathbf{v}_{t,n} = \begin{cases} \nabla \ell(\mathbf{w}_t, \xi_t) & \text{if } \xi_t = \xi_n \\ \mathbf{v}_{t-1,n} & \text{otherwise} \end{cases} \tag{3.44}$$

At the beginning of training, SAGA evaluates the full batch GD in order to initialize the vectors $\mathbf{v}_{0,n} = \nabla \ell(\mathbf{w}_0, \xi_n)$ for all n. Using the gradient $\mathbf{v}_{t,n}$ stored in memory in place of $N-1$ missing stochastic gradients corresponding to the unsampled indices, reduces the variance of the overall gradient. As a result, SAG can achieve an $O(\log(1/\epsilon))$ linear convergence rate. However, this convergence improvement comes at the cost of $O(ND)$ memory required to store previously computed gradients for all the N training samples, and a one-time $O(N)$ cost of initializing $\mathbf{v}_{t,n}$ by evaluating the full gradient. Other, more efficient initialization techniques could be used in practice such as setting the gradients to zero, or only using the gradients that are available in memory until that time.

A drawback of SAG is that the gradient $\frac{1}{N} \sum_{n=1}^{N} \mathbf{v}_{t,n}$ is a biased estimate of the full gradient $\sum_{n=1}^{N} \nabla \ell(\mathbf{w}_t, \xi_n)$. A variant of SAG called SAGA proposed in [4] removes this bias by using the following modified version of the update rule:

$$\mathbf{w}_{t+1} = \mathbf{w}_t - \eta \left(\nabla \ell(\mathbf{w}_t, \xi_t) - \nabla \ell(\mathbf{v}_{t,n}, \xi_t) + \frac{1}{N} \sum_{n=1}^{N} \mathbf{v}_{t,n} \right) \qquad (3.45)$$

By moving the $\frac{1}{N}$ inside the brackets, SAGA ensures that the expected value of the gradient used in every update is equal to the full batch gradient $\sum_{n=1}^{N} \nabla \ell(\mathbf{w}_t, \xi_n)$, but it has a lower variance.

While we present SAG and SAGA for the case where each stochastic gradient $\nabla \ell(\mathbf{w}_t, \xi_t)$ is computed over a single sample from the training dataset, the algorithms and their convergence analysis can easily be extended to the case where $\nabla \ell(\mathbf{w}_t, \xi_t)$ is computed over a mini-batch of b samples chosen uniformly at random with replacement from the training dataset.

3.3.3   Stochastic Variance Reduced Gradient (SVRG)

Unlike SAG and SAGA, the SVRG method proposed in [3] does not require additional memory to store previous gradients. The key idea behind SVRG is to periodically compute the full gradient, and use it to reduce the variance of stochastic gradients in each iteration. SVRG operates in cycles, where each cycle consists of t_0 iterations. At the end of each cycle, it updates $\tilde{\mathbf{w}}$, which denotes the snapshot of the $\mathbf{w}$, that is, $\tilde{\mathbf{w}}$ is set to $\mathbf{w}_t$ if $t \mod t_0 = 0$. After every update to the snapshot, SVRG also computes the full gradient $\nabla F(\tilde{\mathbf{w}})$ at the snapshot $\tilde{\mathbf{w}}$. During the t_0 iterations of the next cycle, SVRG updates the model parameter vector $\mathbf{w}_t$ as follows:

$$\mathbf{w}_{t+1} = \mathbf{w}_t - \eta (\underbrace{\nabla \ell(\mathbf{w}_t, \xi_t) - \nabla \ell(\tilde{\mathbf{w}}, \xi_t) + \nabla F(\tilde{\mathbf{w}})}_{\tilde{g}(\mathbf{w}_t)}, \qquad (3.46)$$

where the gradient $\tilde{g}(\mathbf{w}_t)$ is an unbiased estimate of the full gradient $\nabla F(\mathbf{w}_t)$, however its variance of the gradient is smaller than the stochastic gradient $\nabla \ell(\mathbf{w}_t, \xi_t)$ at the sample chosen in the t-th iteration.

SVRG gives an iteration complexity of $O(\log(\frac{1}{\epsilon}))$, matching that of batch GD. This faster convergence comes at a additional computation cost of periodically computing the full gradient. Amortizing the per-cycle cost across the t_0 iterations in that cycle, the per-iteration computation cost of SVRG is $O(\frac{N}{t_0})$, in contrast to the $O(b)$ of computing a mini-batch gradient in each iteration of mini-batch SGD.

Similar to SAG and SAGA, although we presented SVRG for the case where $\nabla \ell(\mathbf{w}_t, \xi_t)$ is computed over a single training sample, the algorithm and its analysis can easily be extended to the case where $\nabla \ell(\mathbf{w}_t, \xi_t)$ is computed over a mini-batch of b samples chosen uniformly at random with replacement from the training dataset.

Table 3.1 Comparison of the iteration complexity, computation per iteration and memory cost of various SGD algorithms

Algorithm	Iters. to reach ϵ error	Comp. per iter.	Memory
Batch GD	$O\left(\log\left(\frac{1}{\epsilon}\right)\right)$	$O(Nd)$	$O(d)$
SGD	$O\left(\frac{1}{\epsilon}\right)$	$O(1)$	$O(d)$
Mini-batch SGD	$O\left(\frac{1}{\epsilon}\right)$	$O(bd)$	$O(d)$
SAGA	$O\left(\log\left(\frac{1}{\epsilon}\right)\right)$	$O(d)$	$O(Nd)$
SVRG	$O\left(\log\left(\frac{1}{\epsilon}\right)\right)$	$O(\frac{Nd}{t_0})$	$O(d)$

Summary

In this chapter, we showed the convergence analysis of batch GD and mini-batch SGD for strongly convex and Lipschitz smooth functions, and showed that they require $O(\log\frac{1}{\epsilon})$ and $O(\frac{1}{\epsilon})$ iterations respectively to reach an ϵ error. In order to bridge the gap in their iteration complexity, methods such as SAGA and SVRG are use previously computed gradients to reduce the variance of the stochastic gradient used to updated the model parameters in each iteration. A comparison of all these SGD variants is summarized in Table 3.1.

The strong convexity assumption used in the convergence analyses presented in this chapter can be removed to obtain a more general convergence analysis for Lipschitz smooth but non-convex objective functions. When the objective function is non-convex, GD and SGD only guarantee convergence to stationary point (local minimum or saddle point) of the function, rather than the global optimum. Please refer to [1] for these general convergence analyses results and their proofs. The proofs of the convergence of SVRG and SAGA are also omitted here for the purpose of brevity and can be found in [2–4] respectively.

Problems

1. Consider a linear regression problem, where the goal is to minimize the residual sum of squares error $\sum_{i=1}^{N}(y_i - \mathbf{w}^\top\mathbf{x}_i)^2$ for a training dataset $\mathcal{D} = \{(\mathbf{x}_1, y_1), \ldots, (\mathbf{x}_N, y_n)\}$. Implement mini-batch SGD and plot the residual sum of squares (RSS) error versus the number of iterations for different mini-batch sizes for a fixed value of learning rate η. How does the convergence rate and the error floor depend on the mini-batch size?
2. Prove Theorem 3.3 by following steps similar to the proof of Theorem 3.2. Please refer to [1] for a detailed solution.
3. For the same linear regression considered in Problem 1, implement SAG and SAGA and compare their performance with that of SGD (equivalent to mini-batch SGD with $b = 1$).

References

1. L. Bottou, F. E. Curtis, and J. Nocedal, "Optimization methods for large-scale machine learning," *arXiv preprint* arXiv:1606.04838, Feb. 2018.
2. N. L. Roux, M. Schmidt, and F. R. Bach, "A stochastic gradient method with an exponential convergence rate for finite training sets," in *Advances in Neural Information Processing Systems*, 2012, pp. 2663–2671.
3. R. Johnson and T. Zhang, "Accelerating stochastic gradient descent using predictive variance reduction," in *Advances in Neural Information Processing Systems 26*, C. J. C. Burges, L. Bottou, M. Welling, Z. Ghahramani, and K. Q. Weinberger, Eds. Curran Associates, Inc., 2013, pp. 315–323. [Online]. Available: http://papers.nips.cc/paper/4937-accelerating-stochastic-gradient-descent-using-predictive-variance-reduction.pdf
4. A. Defazio, F. Bach, and S. Lacoste-Julien, "SAGA: A fast incremental gradient method with support for non-strongly convex composite objectives," in *Advances in Neural Information Processing Systems 27*, 2014, pp. 1646–1654.

Synchronous SGD and Straggler-Resilient Variants   4

For large training datasets, it can be prohibitively slow to conduct sequential SGD training at a single node, as we described in Chap. 1. Therefore, most practical implementations run SGD in a distributed manner using multiple *worker nodes* that share the task of computing mini-batch gradients. They communicate the gradients with a central server called the *parameter server* [1] that stores the current version of the model. In this chapter, we will introduce two types of distributed training in the parameter server framework, namely data-parallel and model-parallel training. Then the rest of the chapter will focus on the most commonly used distributed SGD algorithm, synchronous SGD. We will analyze its error convergence and runtime per iteration and study the trade-off between these two quantities. Finally, we will introduce straggler-resilient variants that strike a good balance between error and runtime.

4.1   Parameter Server Framework

To understand the need to distribute SGD across multiple computing nodes, let us evaluate the *wallclock runtime* taken to achieve this error for a simple delay model. Suppose the time taken to compute a sample gradient is y and the time taken to update model parameters is denoted by δ. Then for a mini-batch size of b, the runtime per iteration is $(by + \delta)$. From the mini-batch SGD convergence analysis presented in Chap. 3, we know that using a larger mini-batch size b results in a lower error. However, the wallclock runtime increases linearly in b.

Instead of a sequential implementation of mini-batch SGD, if we use multiple worker nodes that compute gradients in parallel, then we can process large mini-batches without increasing the wallclock time per iteration. For example, suppose we consider a network of b workers that compute one sample gradient each in parallel, the runtime per iteration will reduce to $(y + \delta)$. This idea of processing more data per iteration using multiple worker nodes is called *data-parallel distributed training*. It is implemented using a parameter server

39
G. Joshi, *Optimization Algorithms for Distributed Machine Learning*,
Synthesis Lectures on Learning, Networks, and Algorithms,
https://doi.org/10.1007/978-3-031-19067-4_4

that stores the current version of the model **w** and collects gradients from the worker nodes. Each worker nodes stores a replica of the current version of the model and a partition of the training data.

Besides data parallelism, an orthogonal way to reduce the runtime per iteration of mini-batch SGD is to reduce the per-sample gradient computation time y. The gradient computation time is proportional to the size of the gradient vector $g(\mathbf{w})$ is equal to the number of parameters in the model **w** that we want to train. In model-parallel training, the model itself is partitioned into shards and each worker only computes gradients for one of the model partitions. Thus, if we have m workers, the per-sample gradient computation time y reduces to y/m, and with the simple delay model considered above the runtime per iteration becomes $by/m + \delta$.

The data-parallel and model-parallel training paradigms within the parameter server framework were first introduced by [1]. While both are used in practice, model-parallel training is less common and is mainly used training massive models because it requires additional communication to ensure synchronization of the model partitions. In most of this chapter, we will focus on an algorithm called synchronous SGD which is the standard version of data-parallel distributed training used in most industrial implementations.

4.2   Distributed Synchronous SGD Algorithm

The synchronous distributed SGD algorithm operates within the parameter server framework that consists of a central parameter server and m worker nodes, as shown in Fig. 4.1. The training dataset $\mathcal{D}$ is shuffled and split equally into partitions $\mathcal{D}_1, \ldots, \mathcal{D}_m$ which are stored at the m workers. The parameter server stores the latest version of the model parameters **w**. Each iteration of the synchronous SGD algorithm proceeds as follows:

1. The parameter server sends the current version of the parameter vector $\mathbf{w}_t$ to all the m workers.
2. Each worker i computes a gradient $g(\mathbf{w}_t; \xi_i) = \frac{1}{b}\sum_{j=1}^{b} \nabla \ell(\mathbf{w}_t; \xi_{i,j})$ using a mini-batch of b samples drawn uniformly at random with replacement from its dataset partition D_i.

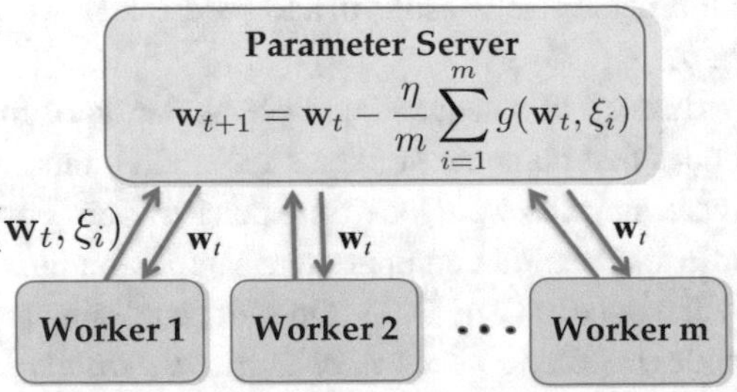

Fig. 4.1 Data-parallel training in the parameter server framework

Parameter Server

$$\mathbf{w}_{t+1} = \mathbf{w}_t - \frac{\eta}{m}\sum_{i=1}^{m} g(\mathbf{w}_t, \xi_i)$$

$g(\mathbf{w}_t, \xi_i)$ $\mathbf{w}_t$ $\mathbf{w}_t$ $\mathbf{w}_t$

Worker 1   Worker 2 $\cdots$ Worker m

3. The parameter server collects these m mini-batch gradients and updates the parameter vector as: $\mathbf{w}_{t+1} = \mathbf{w}_t - \eta \frac{1}{m} \sum_{i=1}^{m} g(\mathbf{w}_t; \xi_i)$.

It is easy to see that synchronous SGD with a mini-batch size b at each of the m workers processes m times more training data per iteration than mini-batch SGD run at the single node. However, does that imply that the error convergence is always better as the number of workers m increases? To answer this questions we adopt our system-aware philosophy that analyzes two dimensions of error convergence, firstly, how the error versus iterations convergence depends on m (in Sect. 4.3 below) and secondly, how the runtime per iteration depends on m (in Sect. 4.4 below). By combining these two factors we can determine how the choice of the number of workers m affects the error versus runtime convergence of distributed synchronous SGD.

4.3 Convergence Analysis

Let us first analysis the error versus iterations convergence of synchronous mini-batch SGD, that is, obtain a bound on $\mathbb{E}\left[F(\mathbf{w}_{t+1})\right] - F(\mathbf{w}^*)$ in terms of the number of workers m, mini-batch size b. Observe that the update rule $\mathbf{w}_{t+1} = \mathbf{w}_t - \eta \frac{1}{m} \sum_{i=1}^{m} g(\mathbf{w}_t; \xi_i)$ of synchronous SGD is equivalent to performing mini-batch SGD, but with a batch size of mb instead of just b samples! So, the convergence analysis of mini-batch SGD can be directly applied here. The only parameter that changes is the variance upper bound.

More formally, the convergence analysis of synchronous SGD with m workers can be obtained as follows. Similar to the analysis mini-batch SGD presented in Chap. 3, we assume that the objective function $F(\mathbf{w})$ is c-strong convex and L-Lipschitz smooth. In addition, we make the following assumptions about the the stochastic gradients $g(\mathbf{w}; \xi)$ returned by the workers:

- **Unbiased Estimate**: The stochastic gradient $g(\mathbf{w}; \xi)$ is an unbiased estimate of $\nabla F(\mathbf{w})$, that is,

$$\mathbb{E}_\xi[g(\mathbf{w}; \xi)] = \nabla F(\mathbf{w})$$

This assumption is true because we consider that in the parameter server framework, the training dataset is *shuffled* and then equally partitioned across the workers.
- **Bounded Variance**: The gradient $\nabla \ell(\mathbf{w}, \xi_l)$ of the l-th sample in each mini-batch has bounded variance, that is,

$$\text{Var}(\nabla \ell(\mathbf{w}; \xi)) \leq \sigma^2 \tag{4.1}$$

This implies the following bounds on the variance of the average stochastic gradient $\frac{1}{m} \sum_{i=1}^{m} g(\mathbf{w}_t; \xi_i)$ used by the parameter server to update the model parameters $\mathbf{w}$:

$$\text{Var}(\frac{1}{m}\sum_{i=1}^{m} g(\mathbf{w}_t; \xi_i)) \leq \frac{\sigma^2}{bm} \tag{4.2}$$

$$\mathbb{E}_{\xi}[\|\frac{1}{m}\sum_{i=1}^{m} g(\mathbf{w}_t; \xi_i)\|^2] \leq \|\nabla F(\mathbf{w})\|^2 + \frac{\sigma^2}{bm} \tag{4.3}$$

Under these assumptions, we have the following result on the convergence of distributed synchronous SGD.

Theorem 4.1 (Convergence of synchronous distributed SGD) *Consider a c-strongly convex and L-smooth function satisfying the unbiased gradient estimate and bounded variance assumptions listed above. if the learning rate $\eta < \frac{1}{L}$ and the starting point is $\mathbf{w}_0$, then $F(\mathbf{w}_t)$ after t synchronous distributed SGD iterations is bounded as*

$$\mathbb{E}[F(\mathbf{w}_t)] - F(\mathbf{w}^*) - \frac{\eta L\sigma^2}{2cbm} \leq (1 - \eta c)^t \left(\mathbb{E}[F(\mathbf{w}_0)] - F(\mathbf{w}^*)) - \frac{\eta L\sigma^2}{2cbm} \right) \tag{4.4}$$

4.3.1 Iteration Complexity

The main implication of Theorem 4.1 above is that the error floor $\frac{\eta L\sigma^2}{2cbm}$ is inversely proportional to the number of workers m and the mini-batch size b at each worker. With a decaying learning rate schedule $\eta_t = \eta_0/t$, to reach an error floor ϵ, synchronous SGD would take $O(1/m\epsilon)$ iterations. Thus, having m times more worker nodes always improves the error versus iterations convergence to a target error by a factor of m iterations.

4.4   Runtime per Iteration

In each iteration of synchronous SGD, the parameter server needs to wait for all m workers to return their gradients, update the parameter vector $\mathbf{w}$, and broadcast it to all the workers. Assuming the ideal scenario where workers perform perfectly parallel computation, the wallclock time per iteration should be independent of m. However, in practice, the workers experience system-level variabilities in their gradient computation times and the communication delay in exchanging gradients and the updated model with the parameter server can also take a non-negligible time. As a result, having m workers would not necessarily result in an m-fold speed-up in the data processed per unit time. In this section, we seek to understand how all the gradient computation and communication times affect the runtime per iteration T_{sync} and its dependence on the number of workers m. Putting this runtime analysis together with the error convergence analysis presented in Sect. 4.3 will shed light on the true convergence speed of synchronous SGD in terms of the error versus wallclock runtime.

4.4.1 Gradient Computation and Communication Time

The time taken by a worker to compute and send its mini-batch gradient to the PS can vary randomly across workers and iterations due to several reasons such as fluctuations in the worker's gradient computation speed, variation in the gradient computation time across mini-batches, and network latency and outages that can affect the time taken for the workers to communicate the gradients to the parameter server. We use the random variable X to denote the total time taken by each worker to compute and send its gradient to the parameter server.

While the probability distribution $F_X(x)$ of the random variable X depends on the computing/network infrastructure and communication and protocols, a useful model is the exponential distribution $\mathsf{Exp}(\lambda)$, which we introduced in Chap. 2. Due to its memoryless property, the exponential distribution is especially well-suited for theoretical analysis and we will use it extensively in the runtime analysis of synchronous SGD and its variants. A practical variant of the exponential distribution is a shifted exponential distribution $\Delta + \mathsf{Exp}(\lambda)$ where Δ represents a constant delay in the communication and local computation, and $\mathsf{Exp}(\lambda)$ captures the random fluctuations in this delay. Moreover, the assumption that the gradient computation times X are i.i.d. across workers and iterations is also required for tractability of the runtime analysis. However, in practice, slow (or fast) workers remain slow (or fast) for multiple iterations. Incorporating such memory can make the runtime analysis more realistic.

4.4.2 Expected Runtime per Iteration

Since the parameter server needs to wait for all m workers to return their gradients, the runtime per iteration T_{sync} of synchronous distributed SGD is the maximum of the gradient computation times $X_1, X_2, \ldots, X_m$ of the m workers, as shown in Fig. 4.2. Due to the synchronous nature of gradient aggregation, fast workers that finish their gradient computation early need to remain idle until the slowest worker returns its gradient to the parameter server. Only then the parameter server can compute the average of the m gradients $\sum_{i=1}^{m} g(\mathbf{w}_t; \xi_i)$, and update the model parameters according to the update rule: $\mathbf{w}_{t+1} = \mathbf{w}_t - \eta \frac{1}{m} \sum_{i=1}^{m} g(\mathbf{w}_t; \xi_i)$. As the number of workers m increases, the probability of at least one of the workers being slow also increases, and therefore the expected runtime per iteration $\mathbb{E}\left[T_{sync}\right]$ increases with m. This effect is referred to as the *tail latency* and has been observed and studied [2] in the context of distributed computations in frameworks such as MapReduce [3].

Let us formally evaluate how $\mathbb{E}\left[T_{sync}\right]$ scales with m for different probability distributions $F_X(x)$. Suppose X_i are independent and identically distributed (i.i.d.) across workers $i = 1, \ldots, m$. Then the expected runtime per iteration $\mathbb{E}\left[T_{sync}\right]$ is given by:

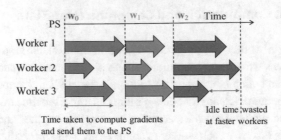

Fig. 4.2 The runtime per iteration of synchronous SGD is the maximum of the gradient computation times (illustrated by the length of each arrow) at each of the m workers. As the number of worker nodes increases, there will be more idle time wasted at the fast workers while the parameter server waits for the slowest workers to finish their gradient computation

$$\mathbb{E}\left[T_{sync}\right] = \mathbb{E}\left[\max(X_1, X_2, \ldots, X_m)\right]$$
$$= \mathbb{E}\left[X_{m:m}\right]$$

where $X_{m:m}$ is the maximum order statistic of the m local gradient computation times.

For example, suppose that the local gradient computation time at each worker $X \sim \Delta + \mathsf{Exp}(\lambda)$, following a shifted exponential distribution, which has mean $\mathbb{E}[X] = \Delta + \frac{1}{\lambda}$. The expected runtime per iteration of synchronous SGD with m workers is:

$$\mathbb{E}\left[T_{sync}\right] = \Delta + \frac{H_m}{\lambda} \sim \Delta + \frac{\log m}{\lambda}$$

where $H_m = \sum_{i=1}^{m} \frac{1}{i}$, the m-th Harmonic number. Thus, the expected runtime increases logarithmically with m.

In Fig. 4.3a, we plot the expected runtime per iteration $\mathbb{E}\left[T_{sync}\right]$ and in Fig. 4.3b, we plot the data processed per iteration, both versus the number of workers, for different distributions of gradient computation time X. All the distributions have the same mean, $\mathbb{E}[X] = 3$, and the ideal scaling plot corresponds to X being constant and equal 3.

In Fig. 4.3a, we observe that with the ideal scaling where each worker takes a constant time X to return its gradient, the expected runtime stays constant irrespective of the number of workers. However, when X is random, as the variance of X increases, the expected runtime grows faster with the number of workers. For example, the shifted exponential distribution $X \sim 2 + \mathsf{Exp}(1)$ has lower variance than the exponential distribution and Pareto distribution with the same mean.

In Fig. 4.3b, we plot the expected number of minibatches processed per unit time, which is equal to $m/\mathbb{E}\left[T_{sync}\right]$. When X is constant, $\mathbb{E}\left[T_{sync}\right] = X$, the data processed per unit time scales linearly with m—the ideal scaling that we desire from parallel computation at the m workers. However, in practice, X is random, due to which we will get a sub-optimal scaling of the data processed per unit time.

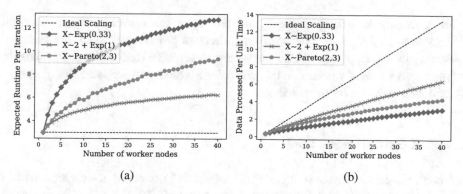

Fig. 4.3 Expected runtime per iteration of synchronous SGD, $\mathbb{E}\left[T_{sync}\right]$, and data processed per unit time versus the number of workers m for different distributions of the gradient computation time X

4.4.3 Error Versus Runtime Convergence

To understand the true convergence speed of distributed synchronous SGD, let us now determine a bound on the error at a wallclock time T by combining the error and runtime analysis given above. Assume that $X \sim \text{Exp}(\lambda)$ following the exponential distribution with rate λ. For large T, the number of iterations t of SGD completed in T units of time is given by

$$\lim_{T \to \infty} \frac{t}{T} = \frac{1}{\mathbb{E}\left[X_{m:m}\right]} \approx \frac{\lambda}{\log m} \qquad (4.5)$$

Thus, if we use $t \approx \frac{T\lambda}{\log m}$ as an approximation for the number of completed iterations, then the error bound at time T is given by:

$$\mathbb{E}\left[F(\mathbf{w}_T)\right] - F(\mathbf{w}^*) - \frac{\eta L \sigma^2}{2cbm} \leq (1 - \eta c)^{\frac{T\lambda}{\log m}} \left(\mathbb{E}\left[F(\mathbf{w}_0)\right] - F(\mathbf{w}^*)) - \frac{\eta L \sigma^2}{2cbm}\right) \qquad (4.6)$$

Observe from this bound that there is a trade-off between the speed of convergence and the error floor. A larger number of workers m results in slower convergence (small exponent of $(1 - \eta c)$) because the expected runtime per iteration is larger, but it achieves a smaller error floor. The wallclock time taken to achieve an error floor ϵ is $O(\frac{\log m}{\lambda \epsilon m})$, since the number of iterations is $O(\frac{1}{\epsilon m})$ and the expected runtime per iteration is $O(\frac{\log m}{\lambda})$.

4.5 Straggler-Resilient Variants

A key takeaway from the runtime analysis of synchronous SGD presented in Sect. 4.4 above is as the number of workers m increases, the tail latency to wait for slowest worker can significantly slow down the error versus runtime convergence. To avoid waiting for straggling workers, we can design straggler-resilient variants of synchronous SGD that modify the gradient aggregation protocol so that the parameter server (PS) only needs to wait for subset of rather than all the m workers. In this section, we present two such variants, proposed in [4], and their error and runtime analyses.

4.5.1 K-Synchronous SGD

The first straggler-resilient variant of synchronous SGD is called K-synchronous SGD, illustrated in Fig. 4.4. In each iteration of K-synchronous SGD:

1. The PS sends the current version of the model $\mathbf{w}_t$ to all m workers.
2. Each worker i compute gradient $g(\mathbf{w}_t; \xi_i)$ using a mini-batch of b samples from this dataset $\mathcal{D}_i$.
3. The parameter server *waits until it receives. gradients from any K of the m workers and discards the rest of the $m - K$ gradients by canceling the gradient computation tasks at the slow workers.* It uses these K gradients to update the model $\mathbf{w}_{t+1} = \mathbf{w}_t - \eta \frac{1}{K} \sum_{i=1}^{K} g(\mathbf{w}_t; \xi_i)$.

Synchronous distributed SGD described in Sect. 4.2 is a special case of K-synchronous SGD when $K = 1$.

Waiting for the $K < m$ fastest workers in each iteration significantly reduces the tail latency and decreases the expected runtime per iteration as compared to m-synchronous SGD. For example, for exponentially distributed gradient computation time $X \sim \text{Exp}(\lambda)$, the expected runtime per iteration is:

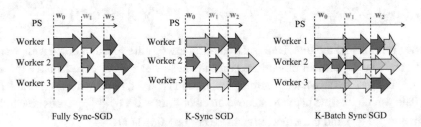

Fig. 4.4 Illustration of synchronous SGD and its straggler-resilient variants K-synchronous SGD and K-batch-synchronous SGD

$$\mathbb{E}\left[T_{K-sync}\right] = \mathbb{E}\left[X_{k:m}\right] \tag{4.7}$$

$$= \frac{1}{m\lambda} + \frac{1}{(m-1)\lambda} + \cdots + \frac{1}{(m-k+1)\lambda} \tag{4.8}$$

$$= \frac{1}{\lambda}\left(H_m - H_{m-K}\right) \tag{4.9}$$

$$\approx \frac{1}{\lambda}\log\left(\frac{m}{m-K}\right). \tag{4.10}$$

The proof follows from the memoryless property of the exponential distribution. The time taken for the fastest worker to finish is the minimum of m exponentials, which is also an exponential with rate $m\lambda$ and mean $1/m\lambda$. Then by the memoryless property, after the fastest worker finishes, the residual time taken by the second fastest worker is also an exponential with rate $(m-1)\lambda$ and mean $1/(m-1)\lambda$. Continuing recursively, we get the expression in (4.9). Then using the approximation $H_m \approx \log m$, we get (4.10). From (4.10) observe that by setting $K = \Theta(m)$ we can cut down on the $\log m$ scaling of the runtime of synchronous SGD.

After receiving the K gradients from the fastest workers, the PS cancels the computation at the other $m - K$ workers. Thus only K mini-batches of data are processed in each iteration, which results in the averaged gradient having a larger variance than in fully synchronous distributed SGD. From the mini-batch SGD convergence analysis, it follows that the error after t iterations of K-synchronous SGD can be bounded as

$$\mathbb{E}\left[F(\mathbf{w}_t)\right] - F(\mathbf{w}^*) - \frac{\eta L\sigma^2}{2cbK} \le (1 - \eta c)^t \left(\mathbb{E}\left[F(\mathbf{w}_0)\right] - F(\mathbf{w}^*) - \frac{\eta L\sigma^2}{2cbK}\right), \tag{4.11}$$

under the same assumptions as in Sect. 4.3. Thus, there is a trade-off between error and runtime as the parameter K changes. A smaller K will reduce the runtime per iteration but it will result in a higher error floor.

4.5.2 K-Batch-Synchronous SGD

While K-synchronous SGD reduces the straggling effect, there is still some idle time at the $K - 1$ workers until the PS waits for the K-th fastest worker to finish its gradient computation. To overcome this residual straggling effect, the K-batch-synchronous SGD variant of synchronous SGD allows fast workers to compute multiple mini-batch gradients per iteration, as illustrated in Fig. 4.4. Each iteration of K-batch-synchronous SGD proceeds as follows:

1. The PS sends the current version of the model $\mathbf{w}_t$ to all m workers.
2. Each worker i continuously computes one or more mini-batch gradients $g(\mathbf{w}_t; \xi_i)$ and send them to the PS, until it receives a cancellation signal and/or the updated $\mathbf{w}_{t+1}$ from the PS.

3. The parameter server waits until it receives K gradients (no matter which worker sends them) in total. It then cancels all outstanding gradient computations at the workers, and updates the model $\mathbf{w}_{t+1} = \mathbf{w}_t - \eta \frac{1}{K} \sum_{i=1}^{K} g(\mathbf{w}_t; \xi_i)$.

For a general probability distribution F_X, it is difficult to obtain a closed-form expression for the expected runtime of K-batch-sync SGD. However, for exponentially distributed gradient computation times $X \sim \text{Exp}(\lambda)$, we can show that the expected runtime per iteration of K-batch-synchronous SGD is given by

$$\mathbb{E}\left[T_{K\text{-}batch\text{-}sync}\right] = \frac{K}{m\lambda}. \tag{4.12}$$

Observe that in contrast to the runtime of synchronous SGD that increases with m, $\mathbb{E}\left[T_{K\text{-}batch\text{-}sync}\right]$ decreases with the number of workers m. Thus, by enabling fast workers to *steal* work from slow workers, we can eliminate the tail latency due to straggling workers and achieve the ideal linear speed-up in the runtime as the m increases.

Under the same assumptions as Sect. 4.3, the error versus iterations convergence is the same as that of K-synchronous SGD given in (4.11). Putting the error and runtime analysis together, we get an error-runtime trade-off controlled by the parameter K, similar to the K-synchronous SGD variant. However, since the error-versus-iterations convergence is identical, but the expected runtime per iteration of K-batch-synchronous is less than of K-synchronous SGD, overall, K-batch-synchronous SGD has a better error versus runtime convergence.

Summary

In this chapter we introduced distributed implementations of SGD using the parameter server framework. Synchronous SGD is the standard approach to perform data-parallel distributed training. We analyzed how its error versus iterations convergence and expected runtime per iteration scale with the number of workers m. A key insight from this analysis is that the runtime per iteration increases with m due to straggling workers. To overcome this tail latency we presented two variants K-synchronous and K-batch-synchronous SGD that relax the synchronization protocol allowing it to only wait for a subset of the workers' gradients in each iteration. By changing the parameter K we can control the degree of synchronization, and the resulting trade-off between error convergence and runtime.

Problems

1. For exponential gradient computation times $X \sim \text{Exp}(\lambda)$, derive the expression for the variance of the runtime per iteration of T_{sync} and analyze how it scales with the number of workers m.
2. By simulating and plotting the expected runtime per iteration of synchronous SGD for Pareto distributed $X \sim \text{Pareto}(x_m, \alpha)$, compare its scaling with m for different values of the shape parameter α.

3. For the simulation set up used above, implement K-sync and K-batch synchronous SGD are compare their expected runtime per iteration versus K.

References

1. J. Dean, G. S. Corrado, R. Monga, K. Chen, M. Devin, Q. V. Le, M. Z. Mao, M. Ranzato, A. Senior, P. Tucker, K. Yang, and A. Y. Ng, "Large scale distributed deep networks," in *Proceedings of the International Conference on Neural Information Processing Systems*, 2012, pp. 1223–1231.
2. J. Dean and L. A. Barroso, "The tail at scale," *Communications of the ACM*, vol. 56, no. 2, pp. 74–80, 2013.
3. J. Dean and S. Ghemawat, "MapReduce: simplified data processing on large clusters," *ACM Commun. Mag.*, vol. 51, no. 1, pp. 107–113, Jan. 2008.
4. S. Dutta, G. Joshi, S. Ghosh, P. Dube, and P. Nagpurkar, "Slow and Stale Gradients Can Win the Race: Error-Runtime Trade-offs in Distributed SGD," *International Conference on Artificial Intelligence and Statistics (AISTATS)*, Apr. 2018. [Online]. Available: https://arxiv.org/abs/1803.01113

Asynchronous SGD and Staleness-Reduced Variants

5

In Chap. 4 we saw that synchronous distributed SGD suffers from tail latency due to straggling workers. Its straggler-resilient variants K-sync and K-batch-sync SGD overcome the straggling delay, but they end up discarding partially completed gradient computation at one or more workers, thus reducing the statistical efficiency of the algorithm and adversely affecting the error convergence.

In this chapter, we study the class of *asynchronous SGD* algorithms that relax the need for all workers to compute gradients for the same synchronized version of the model **w**. First popularized by [1, 2], asynchronous SGD has been extensively studied in recent distributed ML literature [3–8]. Allowing workers to have different versions of the model causes *staleness* in some of the gradients. In Sect. 5.3 we analyze how staleness affects error convergence, and in Sect. 5.2 we analyze the runtime per iteration of asynchronous SGD. To limit the gradient staleness while preserving the runtime benefits of asynchronous SGD, we study some staleness-reduced variants in Sect. 5.4.

5.1   The Asynchronous SGD Algorithm

Similar to synchronous distributed SGD, asynchronous SGD is a data-parallel distributed training algorithm that operates within the parameter server framework. The training dataset $\mathcal{D}$ is shuffled and uniformly partitioned into subsets $\mathcal{D}_1, \ldots, \mathcal{D}_m$ that are stored at worker nodes $1, 2, \ldots, m$ respectively. In asynchronous SGD, each worker i, for $i = 1, \ldots, m$ operates independently and does the following:

© The Author(s), under exclusive license to Springer Nature Switzerland AG 2023
G. Joshi, *Optimization Algorithms for Distributed Machine Learning*,
Synthesis Lectures on Learning, Networks, and Algorithms,
https://doi.org/10.1007/978-3-031-19067-4_5

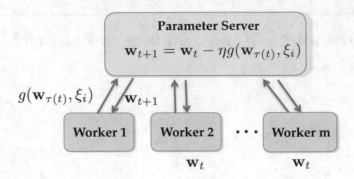

Fig. 5.1 In asynchronous SGD, whenever a worker sends its gradient to the PS, the PS updates the model **w** and send the new version only to that worker. Other workers continue computing gradients at older versions of the model

1. Pull the current version of the model $\mathbf{w}_t$ from the parameter server.
2. Compute the mini-batch gradient $g(\mathbf{w}_t; \xi_i)$ using a mini-batch of data sampled from its local dataset partition $\mathcal{D}_i$.

As soon as the parameter server receives a gradient from one of the workers, it updates the model and sends the new version to that worker. The local gradient computation and communication time can fluctuate across mini-batches and workers. Thus, while worker i is completing its gradient computation, one or more other workers may update the model at the parameter server. As a result, worker i's gradient would be computed at a stale version of the model **w** (Fig. 5.1).

More formally, if $\mathbf{w}_t$ is the current version of the model at the parameter server, it may receive from the worker a (potentially stale) gradient $g(\mathbf{w}_{\tau(t)})$ for some index $\tau(t) \leq t$. Thus, the asynchronous SGD update rule is:

$$\mathbf{w}_{t+1} = \mathbf{w}_t - \eta g(\mathbf{w}_{\tau(t)}; \xi). \tag{5.1}$$

The index $\tau(t)$ is a random variable that depends on the distribution of the gradient computation time X at each worker, and the number of workers.

5.1.1 Comparison with Synchronous SGD

In each iteration of synchronous SGD, the parameter server waits for all m workers to send one mini-batch gradient each. Thus, m mini-batches of b samples each are processed in every iteration. In contrast, in asynchronous SGD, the parameter only waits for one worker at a time to send its gradient and thus it can complete significantly more iterations in the same time. However, only one mini-batch is processed in each iteration and the gradient used to update the model $\mathbf{w}_t$ can also be a stale gradient computed at a previous version $\mathbf{w}_{\tau(t)}$. In the rest of this chapter, we will take a deep dive into the error-runtime trade-off of asynchronous SGD and understand how it compares with synchronous SGD and its variants.

5.2   Runtime Analysis

In synchronous SGD and its variants suffer from inefficiency due to idle time at fast workers and discarded partial computation at slow workers. By removing the synchronization barrier, asynchronous SGD eliminates this runtime inefficiency, albeit at the cost of gradient staleness.

For gradient computation time $X \sim F_X$, the expected runtime per iteration is given by

$$\mathbb{E}\left[T_{async}\right] = \frac{\mathbb{E}\left[X\right]}{m} \tag{5.2}$$

The proof comes from the elementary renewal theorem [9, Chap. 5]. For the i-th worker, let $A_i(t)$ by the number of gradients it pushes to the PS in time t. The time between two consecutive gradient pushes is an independent realization $X_i \sim F_X$, whose mean is $\mathbb{E}\left[X\right]$. By the elementary renewal theorem we have:

$$\lim_{t \to \infty} \frac{A_i(t)}{t} = \frac{1}{\mathbb{E}\left[X\right]} \tag{5.3}$$

Since m workers push gradients independently, the total number of gradient pushes $\sum_{i=1}^{m} A_i(t)$ in time t, which is also equal to the number of iterations completed in time t, is a superposition of m renewal processes and it satisfies:

$$\lim_{t \to \infty} \frac{\sum_{i=1}^{m} A_i(t)}{t} = \frac{m}{\mathbb{E}\left[X\right]} \tag{5.4}$$

Thus expected runtime per iteration is equal to $\frac{\mathbb{E}[X]}{m}$.

5.2.1   Runtime Speed-Up Compared to Synchronous SGD

For exponentially distributed gradient computation times, $X \sim \text{Exp}(\lambda)$ the ratio of the expected runtimes per iteration of synchronous and asynchronous SGD is:

$$\frac{\mathbb{E}\left[T_{sync}\right]}{\mathbb{E}\left[T_{async}\right]} \approx m \log m \tag{5.5}$$

This, asynchronous SGD gives an $O(m \log m)$ runtime speed-up and it can be dramatically faster than synchronous SGD for large m.

5.3   Convergence Analysis

Next, let us analyze how the error versus iterations convergence of asynchronous SGD is affected by the staleness in gradients returned by the worker nodes. The convergence result presented in Theorem 5.1 requires the following assumptions. They are similar to the assumptions that we made in the single-node SGD and distributed synchronous SGD analyses, except that we need an additional bound on the staleness of gradients.

1. **Lipschitz Smoothness**: $F(\mathbf{w})$ is an L-smooth function. Thus,

$$||\nabla F(\mathbf{w}) - \nabla F(\widetilde{\mathbf{w}})||_2 \leq L||\mathbf{w} - \widetilde{\mathbf{w}}||_2 \quad \forall \mathbf{w}, \widetilde{\mathbf{w}}. \tag{5.6}$$

2. **Strong Convexity**: $F(\mathbf{w})$ is strongly convex with parameter c. Thus,

$$2c(F(\mathbf{w}) - F^*) \leq ||\nabla F(\mathbf{w})||_2^2 \quad \forall \mathbf{w}. \tag{5.7}$$

3. **Unbiased Stochastic Gradients**: The stochastic gradient is an unbiased estimate of the true gradient:

$$\mathbb{E}_{\xi_j|\mathbf{w}_k} \left[g(\mathbf{w}_k, \xi_j) \right] = \nabla F(\mathbf{w}_k) \quad \forall k \leq j. \tag{5.8}$$

Observe that this is slightly different from the common assumption that says $\mathbb{E}_{\xi_j} \left[g(\mathbf{w}, \xi_j) \right] = \nabla F(\mathbf{w})$ for all $\mathbf{w}$. Observe that all $\mathbf{w}_j$ for $j > k$ is actually not independent of the data ξ_j. We thus make the assumption more rigorous by conditioning on $\mathbf{w}_k$ for $k \leq j$. Our requirement $k \leq j$ means that $\mathbf{w}_k$ is the value of the parameter at the PS before the data ξ_j was accessed and can thus be assumed to be independent of the data ξ_j.

4. **Bounded Gradient Variance**: We assume that the variance of the stochastic update given $\mathbf{w}_k$ at iteration k before the data point was accessed is also bounded as follows:

$$\mathbb{E}_{\xi_j|\mathbf{w}_k} \left[||g(\mathbf{w}_k, \xi_j) - \nabla F(\mathbf{w}_k)||_2^2 \right] \leq \frac{\sigma^2}{b} \quad \forall k \leq j. \tag{5.9}$$

5. **Bounded Staleness**: We assume that for some $\gamma \leq 1$,

$$\mathbb{E} \left[||\nabla F(\mathbf{w}_j) - \nabla F(\mathbf{w}_{\tau(j)})||_2^2 \right] \leq \gamma \mathbb{E} \left[||\nabla F(\mathbf{w}_j)||_2^2 \right].$$

Recall that $\tau(j)$, the index of the model version at the worker that returns a gradient at iteration j, is a random variable. Its distribution implies a parameter p_0, a lower bound on the probability of receiving a fresh gradient in an iteration, as we show in the Lemma 5.1 below. This parameter will feature in the error bound in Theorem 5.1 below.

Lemma 5.1 *Suppose that* $p_0^{(j)}$ *is the conditional probability that* $\tau(j) = j$ *given all the past delays and all the previous* $\mathbf{w}$, *and* $p_0 \le p_0^{(j)}$ *for all* j. *Then,*

$$\mathbb{E}\left[||\nabla F(\mathbf{w}_{\tau(j)})||_2^2\right] \ge p_0\mathbb{E}\left[||\nabla F(\mathbf{w}_j)||_2^2\right]. \tag{5.10}$$

Proof By the law of total expectation,

$$\mathbb{E}\left[||\nabla F(\mathbf{w}_{\tau(j)})||_2^2\right] = p_0^{(j)}\mathbb{E}\left[||\nabla F(\mathbf{w}_{\tau(j)})||_2^2|\tau(j) = j\right]$$
$$+ (1 - p_0^{(j)})\mathbb{E}\left[||\nabla F(\mathbf{w}_{\tau(j)})||_2^2|\tau(j) \ne j\right]$$
$$\ge p_0\mathbb{E}\left[||\nabla F(\mathbf{w}_j)||_2^2\right].$$

If the gradient computation time X at each worker follows an exponential distribution, then we can show that $\tau(j)$ is geometrically distributed, and $p_0 = \frac{1}{m}$. This is because, when the gradient computation times are exponentially distributed, the number of gradients received by the parameter server within a time window follows the Poisson distribution, and each gradient push is equally likely to come from any of the m worker nodes.

With the above assumptions, we can show the following convergence bound on the error after t iterations of asynchronous SGD.

Theorem 5.1 (Error convergence of asynchronous SGD) *Suppose the objective* $F(\mathbf{w})$ *is* c-*strongly convex and the learning rate* $\eta \le \frac{1}{2L}$. *Also assume that for some* $\gamma \le 1$,

$$\mathbb{E}\left[||\nabla F(\mathbf{w}_j) - \nabla F(\mathbf{w}_{\tau(j)})||_2^2\right] \le \gamma\mathbb{E}\left[||\nabla F(\mathbf{w}_j)||_2^2\right].$$

Then, the error of after t iterations is,

$$\mathbb{E}\left[F(\mathbf{w}_t)\right] - F^* \le \frac{\eta L\sigma^2}{2c\gamma'b} + (1 - \eta c\gamma')^t\left(\mathbb{E}\left[F(\mathbf{w}_0)\right] - F^* - \frac{\eta L\sigma^2}{2c\gamma'b}\right) \tag{5.11}$$

where $\gamma' = 1 - \gamma + \frac{p_0}{2}$.

Proof We start with the L-Lipschitz smoothness assumption, which implies the following:

$$F(\mathbf{w}_{j+1}) \le F(\mathbf{w}_j) + (\mathbf{w}_{j+1} - \mathbf{w}_j)^T\nabla F(\mathbf{w}_j) + \frac{L}{2}||\mathbf{w}_{j+1} - \mathbf{w}_j||_2^2$$
$$= F(\mathbf{w}_j) + (-\eta g(\mathbf{w}_{\tau(j)}))^T\nabla F(\mathbf{w}_j) + \frac{L\eta^2}{2}||g(\mathbf{w}_{\tau(j)})||_2^2$$
$$= F(\mathbf{w}_j) - \frac{\eta}{2}||\nabla F(\mathbf{w}_j)||_2^2 - \frac{\eta}{2}||g(\mathbf{w}_{\tau(j)})||_2^2 + \frac{\eta}{2}||\nabla F(\mathbf{w}_j) - g(\mathbf{w}_{\tau(j)})||_2^2$$
$$+ \frac{L\eta^2}{2}||g(\mathbf{w}_{\tau(j)})||_2^2. \tag{5.12}$$

where the last line follows from $2a^T b = ||a||_2^2 + ||b||_2^2 - ||a - b||_2^2$. Taking expectation on both sides we have

$$\mathbb{E}\left[F(\mathbf{w}_{j+1})\right] - \mathbb{E}\left[F(\mathbf{w}_j)\right] \tag{5.13}$$

$$\leq -\frac{\eta}{2}\mathbb{E}\left[||\nabla F(\mathbf{w}_j)||_2^2\right] - \frac{\eta}{2}\mathbb{E}\left[||g(\mathbf{w}_{\tau(j)})||_2^2\right]$$

$$+ \frac{\eta}{2}\mathbb{E}\left[||\nabla F(\mathbf{w}_j) - g(\mathbf{w}_{\tau(j)})||_2^2\right] + \frac{L\eta^2}{2}\mathbb{E}\left[||g(\mathbf{w}_{\tau(j)})||_2^2\right] \tag{5.14}$$

$$\leq -\frac{\eta}{2}\mathbb{E}\left[||\nabla F(\mathbf{w}_j)||_2^2\right] - \frac{\eta}{2}\mathbb{E}\left[||g(\mathbf{w}_{\tau(j)})||_2^2\right] + \frac{L\eta^2}{2}\mathbb{E}\left[||g(\mathbf{w}_{\tau(j)})||_2^2\right]$$

$$+ \frac{\eta}{2}\mathbb{E}\left[||\nabla F(\mathbf{w}_j) - \nabla F(\mathbf{w}_{\tau(j)}) + \nabla F(\mathbf{w}_{\tau(j)}) - g(\mathbf{w}_{\tau(j)})||_2^2\right] \tag{5.15}$$

$$\leq -\frac{\eta}{2}\mathbb{E}\left[||\nabla F(\mathbf{w}_j)||_2^2\right] - \frac{\eta}{2}\mathbb{E}\left[||g(\mathbf{w}_{\tau(j)})||_2^2\right] + \frac{L\eta^2}{2}\mathbb{E}\left[||g(\mathbf{w}_{\tau(j)})||_2^2\right]$$

$$+ \frac{\eta}{2}\mathbb{E}\left[||\nabla F(\mathbf{w}_j) - \nabla F(\mathbf{w}_{\tau(j)})||_2^2\right] + \frac{\eta}{2}\mathbb{E}\left[||\nabla F(\mathbf{w}_{\tau(j)}) - g(\mathbf{w}_{\tau(j)})||_2^2\right] \tag{5.16}$$

$$\leq -\frac{\eta}{2}\mathbb{E}\left[||\nabla F(\mathbf{w}_j)||_2^2\right] - \frac{\eta}{2}\mathbb{E}\left[||g(\mathbf{w}_{\tau(j)})||_2^2\right] + \frac{L\eta^2}{2}\mathbb{E}\left[||g(\mathbf{w}_{\tau(j)})||_2^2\right]$$

$$+ \frac{\eta}{2}\mathbb{E}\left[||\nabla F(\mathbf{w}_j) - \nabla F(\mathbf{w}_{\tau(j)})||_2^2\right] + \frac{\eta}{2}\mathbb{E}\left[||g(\mathbf{w}_{\tau(j)})||_2^2\right] - \frac{\eta}{2}\mathbb{E}\left[||\nabla F(\mathbf{w}_{\tau(j)})||_2^2\right] \tag{5.17}$$

$$\leq -\frac{\eta}{2}\mathbb{E}\left[||\nabla F(\mathbf{w}_j)||_2^2\right] - \frac{\eta}{2}\mathbb{E}\left[||\nabla F(\mathbf{w}_{\tau(j)})||_2^2\right] + \frac{L\eta^2}{2}\mathbb{E}\left[||g(\mathbf{w}_{\tau(j)})||_2^2\right]$$

$$+ \frac{\eta}{2}\mathbb{E}\left[||\nabla F(\mathbf{w}_j) - \nabla F(\mathbf{w}_{\tau(j)})||_2^2\right] \tag{5.18}$$

where

- in (5.15) we add and subtract the full (potentially) stale gradient $\nabla F(\mathbf{w}_{\tau(j)})$ in the term with the difference of full fresh gradient $\nabla F(\mathbf{w}_j)$ and the stochastic (potentially) stale gradient $g(\mathbf{w}_{\tau(j)})$,
- in (5.16) we open the norm-squared of the summation, and use the fact that the cross term is zero because $\mathbb{E}\left[\nabla F(\mathbf{w}_{\tau(j)}) - g(\mathbf{w}_{\tau(j)})\right] = 0$ as a result of the unbiased gradient assumption,
- to get (5.17) we further expand the last term in (5.16) and use the unbiased gradient assumption, which implies that the cross term $-2\nabla F(\mathbf{w}_{\tau(j)})^\top \mathbb{E}\left[g(\mathbf{w}_{\tau(j)})\right] = -2||\nabla F(\mathbf{w}_{\tau(j)})||_2^2$, and
- finally, to get (5.18) we cancel out the two terms $-\frac{\eta}{2}\mathbb{E}\left[||g(\mathbf{w}_{\tau(j)})||_2^2\right]$ and $\frac{\eta}{2}\mathbb{E}\left[||g(\mathbf{w}_{\tau(j)})||_2^2\right]$. $\qquad\square$

Next we use the bounded variance and bounded staleness assumptions to further simplify the right hand side as follows

$$\mathbb{E}\left[F(\mathbf{w}_{j+1})\right] - \mathbb{E}\left[F(\mathbf{w}_j)\right] \tag{5.19}$$

$$\leq -\frac{\eta}{2}(1-\gamma)\mathbb{E}\left[||\nabla F(\mathbf{w}_j)||_2^2\right] + \frac{L\eta^2\sigma^2}{2b} - \frac{\eta}{2}(1-\eta L)\mathbb{E}\left[||\nabla F(\mathbf{w}_{\tau(j)})||_2^2\right] \tag{5.20}$$

$$\leq -\frac{\eta}{2}(1-\gamma)\mathbb{E}\left[||\nabla F(\mathbf{w}_j)||_2^2\right] + \frac{L\eta^2\sigma^2}{2m} - \frac{\eta}{4}p_0\mathbb{E}\left[||\nabla F(\mathbf{w}_j)||_2^2\right]. \tag{5.21}$$

where in (5.21) we use Lemma 5.1.

Next, we will use the assumption that the objective function $F(\mathbf{w})$ c-strongly convex, which implies that

$$2c(F(\mathbf{w}) - F^*) \leq ||\nabla F(\mathbf{w})||_2^2 \quad \forall \mathbf{w} \tag{5.22}$$

Using this result in (5.21), we obtain the following:

$$\mathbb{E}\left[F(\mathbf{w}_{j+1})\right] - F^* \leq \frac{\eta^2 L\sigma^2}{2b} + (1 - \eta c(1 - \gamma + \frac{p_0}{2}))(\mathrm{e}F(\mathbf{w}_j) - F^*).$$

Let us denote $\gamma' = (1 - \gamma + \frac{p_0}{2})$. Then, from the above recursion we get

$$\mathbb{E}[F(\mathbf{w}_t)] - F^* \leq \frac{\eta L\sigma^2}{2c\gamma'b} + (1 - \eta\gamma'c)^t\left(\mathbb{E}[F(\mathbf{w}_0)] - F^* - \frac{\eta L\sigma^2}{2c\gamma'b}\right).$$

5.3.1   Implications of the Asynchronous SGD Convergence Bound

Recall that γ is a measure of staleness of the gradients, with larger γ indicating more staleness. Similarly, smaller p_o also indicates more staleness because the probability of getting a fresh gradient is lower bounded by a smaller quantity. The convergence bound in (5.11) shows the effect of staleness on the convergence speed and the error floor. The parameter $\gamma' = (1 - \gamma + \frac{p_0}{2})$ is smaller when there is higher staleness. Smaller γ' will result in slower convergence speed, that is, larger $(1 - \eta\gamma'c)$, and a higher error floor, that is, larger $\frac{\eta L\sigma^2}{2c\gamma'b}$. If we use an appropriate $O(1/t)$ learning rate decay, then the overall error convergence rate is $O(1/t)$, that is, it takes $O(1/\epsilon)$ to reach an error floor of ϵ.

Although the error versus iterations convergence of asynchronous SGD may be slower than that of synchronous SGD, because the runtime per iteration is significantly smaller, the error versus wallclock runtime convergence may still favor asynchronous SGD, as shown in Fig. 5.2.

5.4   Staleness-Reduced Variants of Asynchronous SGD

Synchronous SGD and asynchronous SGD are at two extremes of the error-runtime trade-off. Synchronous SGD has a lower error floor at convergence but suffers from tail latency and has a higher runtime per iteration. Asynchronous SGD sharply reduces the runtime

Fig. 5.2 Theoretical
error-runtime trade-off for
synchronous and asynchronous
SGD with same η.
Asynchronous SGD has faster
decay with time but a higher
error floor

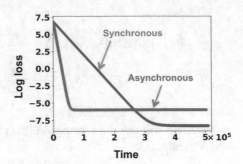

per iteration by removing the synchronization barrier at the PS, but it results in a higher
error floor due to staleness in the gradients returned by workers to the PS. In this section,
we propose two staleness-reduced variants of asynchronous SGD that span the spectrum
between asynchronous and synchronous SGD.

5.4.1 K-Asynchronous SGD

The first staleness-reduced variant of asynchronous SGD is called K-asynchronous SGD,
illustrated in Fig. 5.3, and suggested in [5, 7]. Similar to asynchronous SGD, each worker i,
for $i = 1, \ldots, m$ operates independently. After receiving the current version of the model $\mathbf{w}_t$
from the parameter server, it computed the mini-batch gradient $g(\mathbf{w}_t; \xi_i)$ using a mini-batch
of data sampled from its local dataset partition $\mathcal{D}_i$ and sends it to the parameter server. The
PS waits for any K out of m workers to send gradients and updates the model $\mathbf{w}_i$. However,
it does not cancel the gradient computation at the remaining $m - K$ workers. As a result,
for every update the gradients returned by each worker might be computed at a stale or older
value of the parameter $\mathbf{w}$. The update rule is thus given by:

$$\mathbf{w}_{j+1} = \mathbf{w}_j - \frac{\eta}{K} \sum_{k=1}^{K} g(\mathbf{w}_{\tau(k,j)}, \xi_{k,j}). \tag{5.23}$$

Here $k = 1, 2, \ldots, K$ denotes the index of one of the K workers that contribute to the update
at the corresponding iteration, $\xi_{l,j}$ is one mini-batch of m samples used by the k-th worker at
the j-th iteration and $\tau(k, j)$ denotes the iteration index when the k-th worker last read from
the central PS where $\tau(k, j) \leq j$. Also, $g(\mathbf{w}_{\tau(k,j)}, \xi_{k,j}) = \frac{1}{m} \sum_{\xi \in \xi_{k,j}} \nabla f(\mathbf{w}_{\tau(k,j)}, \xi_{k,j})$ is
the average gradient of the loss function evaluated over the mini-batch $\xi_{k,j}$ based on the
stale value of the parameter $\mathbf{w}_{\tau(k,j)}$. Asynchronous SGD is a special case of K-asynchronous
SGD when $K = 1$, and synchronous SGD is a special case when $K = m$.

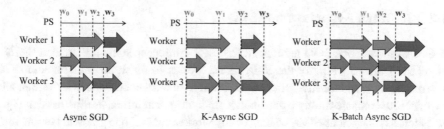

Fig. 5.3 Asynchronous SGD and its staleness-reduced variants K-asynchronous and K-batch-asynchronous SGD for $K = 2$ and $m = 3$

5.4.1.1 Error Convergence of K-Asynchronous SGD

The convergence analysis of K-asynchronous SGD is similar to asynchronous SGD, except that the staleness parameter γ is smaller for larger K, and the mini-batch size is Kb instead of b. Thus, after t iterations, the error of K-asynchronous SGD is bounded as:

$$\mathbb{E}\left[F(\mathbf{w}_t)\right] - F^* \leq \frac{\eta L \sigma^2}{2c\gamma' Kb} + (1 - \eta c\gamma')^t \left(\mathbb{E}\left[F(\mathbf{w}_0)\right] - F^* - \frac{\eta L \sigma^2}{2c\gamma' Kb}\right) \tag{5.24}$$

where $\gamma' = 1 - \gamma + \frac{p_0}{2}$. The proof is a generalized of the proof of Theorem 5.1, and it can be found in [7]. From (5.24), we can see that the error floor can be reduced by increasing the value of K without affecting the convergence speed. However, increasing K can increase the expected runtime per iteration, as we see below.

5.4.1.2 Runtime per Iteration of K-Asynchronous SGD

Recall from (5.2) that the expected runtime per iteration of asynchronous SGD is $\mathbb{E}\left[T_{async}\right] = \mathbb{E}[X]/m$. The parameter K controls the number of gradients that the PS waits for before updating the model $\mathbf{w}$. Increasing K increases the runtime per iteration. While it is difficult to find a closed-form expression for a general distribution F_X, we can evaluate it for exponential distributed gradient computation times $X \sim \mathsf{Exp}(\lambda)$. Due to the memoryless property of the exponential distribution, starting from the beginning of each iteration, the time taken by each worker to finish is $X \sim \mathsf{Exp}(\lambda)$, irrespective of whether the worker is computing a fresh gradient on the updated model or a stale gradient on a previous model version. Thus,

$$\mathbb{E}\left[T_{K\text{-}async}\right] = \mathbb{E}[X_{K:m}] = \frac{H_m - H_K}{\lambda} \tag{5.25}$$

where $X_{K:m}$ is the Kth order statistic of m i.i.d. random variables $X_1, X_2, \ldots, X_m$. We can also show that if X has a new-longer-than-used distribution then, the expected runtime per iteration for K-async is upper-bounded as $\mathbb{E}[T] \leq \mathbb{E}[X_{K:m}]$.

5.4.2 K-Batch-Asynchronous SGD

While K-asynchronous SGD reduces staleness, it introduces some idle time at the $K - 1$ workers until the PS waits for the K-th fastest worker to finish its gradient computation. To overcome this residual straggling effect, the K-batch-asynchronous SGD variant allows fast workers to compute multiple mini-batch gradients per iteration, as illustrated in Fig. 5.3. In K-batch-asynchronous SGD, whenever any worker finishes one gradient computation, it pushes that gradient to the PS, fetches current parameter at PS and starts computing gradient on the next mini-batch. The PS waits for K mini-batch gradients before updating itself but irrespective of which worker they come from. The update rule is similar to Eq. (5.23) theoretically except that now k denotes the indices of the K mini-batches that finish first instead of the workers and $\mathbf{w}_{\tau(k,j)}$ denotes the version of the parameter when the worker computing the k-th mini-batch last read from the PS.

While the error convergence of K-batch-async is similar to K-async, it reduces the runtime per iteration as no worker is idle.

Lemma 5.2 (Runtime of K-batch-async SGD) *The expected runtime per iteration for K-batch-async SGD in the limit of large number of iterations is given by:*

$$\mathbb{E}[T] = \frac{K\mathbb{E}[X]}{m}. \tag{5.26}$$

Proof (Proof of Lemma 5.2) For the i-th worker, let $\{N_i(t), t > 0\}$ be the number of times the i-th worker pushes its gradient to the PS over in time t. The time between two pushes is an independent realization of X_i. Thus, the inter-arrival times $X_i^{(1)}, X_i^{(2)}, \ldots$ are i.i.d. with mean inter-arrival time $\mathbb{E}[X_i]$. Using the elementary renewal theorem [9, Chap. 5] we have,

$$\lim_{t \to \infty} \frac{\mathbb{E}[N_i(t)]}{t} = \frac{1}{\mathbb{E}[X_i]}. \tag{5.27}$$

Thus, the rate of gradient pushes by the i-th worker is $1/\mathbb{E}[X_i]$. As there are m workers, we have a superposition of P renewal processes and thus the average rate of gradient pushes to the PS is

$$\lim_{t \to \infty} \sum_{i=1}^{m} \frac{\mathbb{E}[N_i(t)]}{t} = \sum_{i=1}^{m} \frac{1}{\mathbb{E}[X_i]} = \frac{m}{\mathbb{E}[X]}. \tag{5.28}$$

Every K pushes are one iteration. Thus, the expected runtime per iteration or effectively the expected time for K pushes is given by $\mathbb{E}[T] = \frac{K\mathbb{E}[X]}{m}$.

Fig. 5.4 Asynchronous SGD
and its staleness-reduced
variants K-asynchronous and
K-batch-asynchronous SGD
for $K = 2$ and $m = 3$

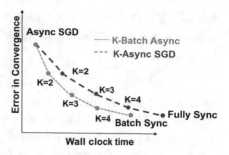

5.5   Adaptive Methods to Improve the Error-Runtime Trade-Off

Together the K-asynchronous and K-batch-asynchronous variants of asynchronous SGD span different points on the trade-off between the error floor at convergence and the wallclock runtime spent iteration, as illustrated in Fig. 5.4. The trade-off is always better for the K-batch asynchronous variants because they eliminate the idle time at the fast workers. By choosing a suitable value of K, we can balance the error floor and runtime.

5.5.1   Adaptive Synchronization

What if we do not have to keep K constant throughout the training process? By dynamically adapting K during training, we can potentially achieve the best of both worlds, a low error floor as well as a fast runtime per iteration. The paper [8] proposes AdaSync, a method that uses the theoretical characterization of the error-runtime trade-off to decide how to gradually increase K during the course of training. As a result, we gradually transition from fully asynchronous SGD ($K = 1$) to fully synchronous SGD ($K = m$). A similar method called Adacomm is proposed for local-update SGD in Chap. 6.

To obtain a schedule for K, let us partition the training time into intervals of time T_0 each. The number of iterations performed within time t is assumed to be approximately $N(t) \approx t/\mathbb{E}[T]$ where $\mathbb{E}[T]$ is the expected runtime per-iteration for the chosen SGD variant. We can write a (heuristic) upper bound on the average of $\mathbb{E}\left[\|\nabla F(\mathbf{w})\|_2^2\right]$ within each time interval T_0 as follows:

$$u(K) = \frac{2(F(\mathbf{w}_{start}) - F^*)\mathbb{E}[T]}{T_0 \eta \gamma'} + \frac{L\eta\sigma^2}{Km\gamma'},$$

where $\mathbf{w}_{start}$ denotes the value of the model $\mathbf{w}$ at the beginning of that time interval. Our goal is to minimize $u(K)$ with respect to K for each time interval. Observe that,

$$\frac{\partial u(K)}{\partial K} = \frac{2(F(\mathbf{w}_{start}))}{T_0 \eta \gamma'} \frac{\partial \mathbb{E}[T]}{\partial K} - \frac{L\eta\sigma^2}{K^2 b\gamma'}. \tag{5.29}$$

Setting $\frac{\partial u(K)}{\partial K}$ to 0 therefore provides a rough heuristic on how to choose parameter K for each time interval, as long as, $\frac{\partial^2 u(K)}{\partial K^2}$ is positive. We approximate $\mathbb{E}[T] \approx \frac{K\Delta}{m} + \frac{K\log m}{m\mu}$. This leads to

$$K^2 = \frac{L\eta\sigma^2 T_0 \eta m \mu}{2m(F(\mathbf{w}_{start}))(\Delta\mu + \log m)}. \tag{5.30}$$

To actually solve for this equation, we would need the values of the Lipschitz constant, variance of the gradient etc. which are not always available. We sidestep this issue by proposing a heuristic here that relies on the ratio of the parameter K at different instants. For K-async SGD, the schedule takes the following form:

$$K \sim K_0 \sqrt{\frac{F(\mathbf{w}_0)}{F(\mathbf{w}_t)}} \tag{5.31}$$

where the initial value K_0 is optimized by grid search.

We evaluate the effectiveness of AdaSync for both K-sync SGD and K-async SGD algorithms. An exponential delay with mean 0.02 s is added to each worker node independently. We fix K for every $t = 60$ s (about 10 epochs). The initial values of K are fine-tuned and set to 2 and 4 for K-sync SGD and K-async SGD, respectively. As shown in Fig. 5.5, the adaptive strategy achieves the fastest convergence in terms of error-versus-time. The adaptive K-async algorithm can even achieve a better error-runtime trade-off than the $K = 8$ case (i.e., fully synchronous case).

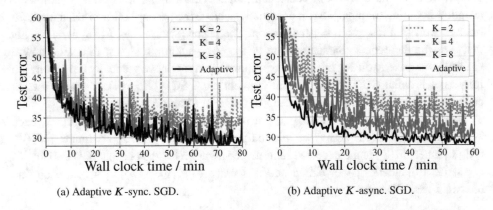

(a) Adaptive K-sync. SGD.   (b) Adaptive K-async. SGD.

Fig. 5.5 Test error of AdaSync SGD on CIFAR-10 with 8 worker nodes. We add an exponential delay with mean 0.02 s on each worker. The value of K is changed after every 60 s

5.5.2   Adaptive Learning Rate Schedule to Compensate Staleness

The staleness of the gradient is random, and can vary across iterations. Intuitively, if the gradient is less stale, we want to weigh it more while updating the parameter $\mathbf{w}$, and if it is more stale we want to scale down its contribution to the update. With this motivation, we propose the following condition on the learning rate at different iterations.

$$\eta_j \mathbb{E}\left[||\mathbf{w}_j - \mathbf{w}_{\tau(j)}||_2^2\right] \leq C \tag{5.32}$$

for a constant C. In our analysis of Asynchronous SGD, we observe that the term $\frac{\eta}{2}\mathbb{E}\left[||\nabla F(\mathbf{w}_j) - \nabla F(\mathbf{w}_{\tau(j)})||_2^2\right]$ is the most difficult to bound. For fixed learning rate, we had assumed that $\mathbb{E}\left[||\nabla F(\mathbf{w}_j) - \nabla F(\mathbf{w}_{\tau(j)})||_2^2\right]$ is bounded by $\gamma ||\nabla F(\mathbf{w}_j)||_2^2$. However, if we impose the condition Eq. (5.32) on η, we do not require this assumption. Our proposed condition actually provides a bound for the staleness term as follows:

$$\frac{\eta_j}{2}\mathbb{E}\left[||\nabla F(\mathbf{w}_j) - \nabla F(\mathbf{w}_{\tau(j)})||_2^2\right] \leq \frac{\eta_j L^2}{2}\mathbb{E}\left[||\mathbf{w}_j - \mathbf{w}_{\tau(j)}||_2^2\right] \leq \frac{CL^2}{2}. \tag{5.33}$$

Inspired by this analysis, we propose the learning rate schedule,

$$\eta_j = \min\left\{\frac{C}{||\mathbf{w}_j - \mathbf{w}_{\tau(j)}||_2^2}, \eta_{max}\right\} \tag{5.34}$$

where η_{max} is a suitably large ceiling on learning rate. It ensures stability when the first term in (5.34) becomes large due to the staleness $||\mathbf{w}_j - \mathbf{w}_{\tau(j)}||_2$ being small. The C is chosen of the same order as the desired error floor. To implement this schedule, the PS needs to store the last read model parameters for every learner. In Fig. 5.6 we illustrate how this schedule can stabilize asynchronous SGD. We also show simulation results that characterize the performance of this algorithm in comparison with naive asynchronous SGD with fixed learning rate.

The idea of variable learning rate is related to the idea of momentum tuning in [4, 10] and may have a similar effect of stabilizing the convergence of asynchronous SGD. However, learning rate tuning is arguably more general since asynchrony results in a momentum term

Fig. 5.6 Adaptive learning rate schedule

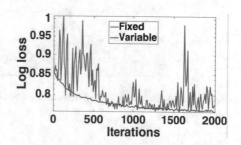

in the gradient update (as shown in [4, 10]) only under the assumption that the staleness process is geometric and independent of **w**.

5.6   HogWild and Lock-Free Parallelism

In the asynchronous SGD algorithms that we have considered in this chapter so far, the entire parameter vector **w** is updated at the same time using gradients from one or more workers. However, in some applications, the objective function $F(\mathbf{w})$ may be sum of sparse functions that only depend on a subset of parameters each, that is,

$$F(\mathbf{w}) = \sum_{e \in \mathcal{E}} f_e(\mathbf{w}_e) \tag{5.35}$$

where e represents a subset of the indices $\{1, 2, \ldots, d\}$ of the parameter vector **w**. The function induces a hypergraph (a graph where an edge can connect more than two vertices) where each hyperedge e connects the set of vertices $\mathbf{w}_e$. For example, each hyperedge is represented by a different color in the illustration of the sparse functions in Fig. 5.7. Some practical applications where the objective function takes this sparse form include sparse support vector machines, matrix completion and graph-cut problems.

The paper [2] proposes an algorithm called HogWild that lets workers update different parts of **w** in a lock-free and asynchronous fashion. The parameter vector is stored in shared memory and it can be accessed by all the workers. Each worker asynchronously does the following:

1. Sample e uniformly at random from the edge set $\mathcal{E}$.
2. Read the subset of indices $\mathbf{w}_e$ from the parameter server and evaluate the gradient $g_e(\mathbf{w}_e) = \nabla f_e(\mathbf{w}_e)$.
3. Update the indicates $\mathbf{w}_e \leftarrow \mathbf{w}_e - \eta g(\mathbf{w}_e)$.

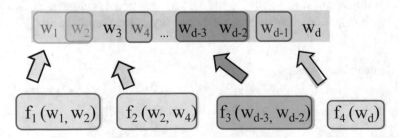

Fig. 5.7 Illustration of the sparse components of the objective function $F(\mathbf{w})$

Observe that depending on the relative speeds of the worker nodes, different parts of the parameter vector $\mathbf{w}$ can have different levels of staleness. Suppose a worker reads a parameter $\mathbf{w}_i$, then by the time it finishes its gradient computation and updates the parameter, one or more other workers may have already updated $\mathbf{w}_i$ to get $\mathbf{w}_j$ for some $j > i$. Thus, the gradient sent by that worker may be $g(\mathbf{w}_{\tau(j)})$ where $\tau(j) < j$ is the index of weight at which that gradient is computed and $j - \tau(j)$ measures the amount of staleness. To prove the convergence of this algorithm, the authors assume that the staleness $j - \tau(j)$ of parameter vector is bounded above by a value B. Under these assumptions, for c-strongly convex and L-Lipschitz smooth functions, they can show an $O(1/t)$ rate of convergence, similar to the rate achieved by asynchronous SGD in Theorem 5.1. The constants in the convergence upper bound depend on the sparsity of the hypergraph induced by the objective function defined in (5.35).

Summary

In this chapter we considered asynchronous SGD, which relaxes the synchronization barrier in synchronous SGD and allows the PS to move forward and update the model without waiting for all workers. Unlike the straggler-resilient variants of synchronous SGD, asynchronous SGD does not cancel the gradient computations at slow workers and instead allows them to send gradients computed at outdated versions of the model. While this reduces wasted computation at the workers, it can introduce staleness in the gradients that are used to update the model. We analyzed the convergence of asynchronous SGD by accounting for this staleness, to show that it can result in a higher error floor than synchronous SGD. We proposed staleness-reduced variants, K-asynchronous SGD and K-batch-asynchronous SGD, which span different points on the trade-off between the error floor and runtime per iteration. Finally, we proposed two adaptive method to overcome staleness, an adaptive synchronization method that dynamically adjusts K during the course of training, and a learning rate schedule that mitigates the adverse effect of stale gradients by reducing the gradient applied to them.

Synchronous and asynchronous SGD are at the backbone of industrial ML implementations today. One of the earliest industrial implementations is Google's DistBelief [11]. Distbelief popularized the data-parallel and model-parallel parameter server framework. Their algorithm Downpour SGD is based on asynchronous SGD. More recently, specialized hardware such as Tensor Processing Units (TPUs) removes straggling and synchronization delays. Thus, industrial implementations are moving from asynchronous SGD back to synchronous SGD. But asynchronous SGD is still a useful paradigm when one is using cheap consumer hardware and low-bandwidth networks.

Problems

1. For exponential gradient computation times $X \sim \text{Exp}(\lambda)$, derive the expression for the expectation and variance of the runtime per iteration of T_{async} and analyze how it scales with the number of workers m.
2. By simulating and plotting the expected runtime per iteration of asynchronous SGD for Pareto distributed $X \sim \text{Pareto}(x_m, \alpha)$, compare its scaling with m for different values of the shape parameter α.
3. For the simulation set up used above, implement K-async and K-batch-async SGD are compare their expected runtime per iteration versus K.

References

1. J. Tsitsiklis, D. Bertsekas, and M. Athans, "Distributed asynchronous deterministic and stochastic gradient optimization algorithms," *IEEE Transactions on Automatic Control*, vol. 31, no. 9, pp. 803–812, 1986.
2. B. Recht, C. Re, S. Wright, and F. Niu, "Hogwild: A lock-free approach to parallelizing stochastic gradient descent," in *Proceedings of the International Conference on Neural Information Processing Systems*, 2011, pp. 693–701.
3. X. Lian, Y. Huang, Y. Li, and J. Liu, "Asynchronous parallel stochastic gradient for nonconvex optimization," in *Proceedings of the International Conference on Neural Information Processing Systems*, 2015, pp. 2737–2745.
4. I. Mitliagkas, C. Zhang, S. Hadjis, and C. Ré, "Asynchrony begets momentum, with an application to deep learning," in *54th Annual Allerton Conference on Communication, Control, and Computing (Allerton)*. IEEE, 2016, pp. 997–1004.
5. S. Gupta, W. Zhang, and F. Wang, "Model accuracy and runtime tradeoff in distributed deep learning: A systematic study," in *IEEE International Conference on Data Mining (ICDM)*. IEEE, 2016, pp. 171–180.
6. X. Lian, W. Zhang, C. Zhang, and J. Liu, "Asynchronous decentralized parallel stochastic gradient descent," in *Proceedings of the 35th International Conference on Machine Learning*, ser. Proceedings of Machine Learning Research, vol. 80. PMLR, 10–15 Jul 2018, pp. 3043–3052. [Online]. Available: http://proceedings.mlr.press/v80/lian18a.html
7. S. Dutta, G. Joshi, S. Ghosh, P. Dube, and P. Nagpurkar, "Slow and Stale Gradients Can Win the Race: Error-Runtime Trade-offs in Distributed SGD," *International Conference on Artificial Intelligence and Statistics (AISTATS)*, Apr. 2018. [Online]. Available: https://arxiv.org/abs/1803.01113
8. S. Dutta, J. Wang, and G. Joshi, "Slow and Stale Gradients Can Win the Race," *IEEE Journal on Selected Areas of Information Theory (JSAIT)*, Aug. 2021.
9. R. Gallager, *Stochastic Processes: Theory for Applications*, 1st ed. Cambridge University Press, 2013.
10. J. Zhang, I. Mitliagkas, and C. Re, "Yellowfin and the art of momentum tuning," *CoRR*, vol. arXiv:1706.03471, Jun. 2017. [Online]. Available: http://arxiv.org/abs/1706.03471
11. J. Dean, G. S. Corrado, R. Monga, K. Chen, M. Devin, Q. V. Le, M. Z. Mao, M. Ranzato, A. Senior, P. Tucker, K. Yang, and A. Y. Ng, "Large scale distributed deep networks," in *Proceedings of the International Conference on Neural Information Processing Systems*, 2012, pp. 1223–1231.

Local-Update and Overlap SGD

6

In the last two chapters, we studied synchronous and asynchronous SGD, where the workers compute mini-batch gradients and the parameter server aggregates them and updates the model. Both these algorithms and their variants require constant communication between the workers and the parameter server after every iteration. However, when the network is bandwidth-limited and/or the size of the model is large, this communication delay can dominate the runtime per iteration. This calls for *communication-efficient* distributed SGD algorithms. In the upcoming chapters, we will study different methods of achieving communication efficiency. In this chapter, we reduce the frequency of communication by allowing workers to performing multiple SGD updates instead of just computing gradients. In Chap. 7, we will study quantization and sparsification methods to reduce the number of bits transmitted in every communication round, and in Chap. 8 we consider decentralized topologies in order to reduce the communication bottleneck at a central parameter server.

6.1   Local-Update SGD Algorithm

Local-update SGD uses a simple strategy to reduce the communication cost of distributed SGD. In local-update SGD, after receiving the latest model parameters **w** from the parameter server, instead of just computing and sending a mini-batch gradient, each worker node makes $\tau > 1$ local updates to the model and then sends the resulting local version of the model back to the parameter server. This strategy reduces the frequency of communication between the parameter server and the workers. The local update SGD algorithm operates in 'communication rounds' or 'training rounds'. More formally, each round proceeds as follows:

1. Worker i fetches the current version of the model $\mathbf{w}_t$ from the parameter server
2. It performs τ local SGD updates using its local data D_i

© The Author(s), under exclusive license to Springer Nature Switzerland AG 2023
G. Joshi, *Optimization Algorithms for Distributed Machine Learning*,
Synthesis Lectures on Learning, Networks, and Algorithms,
https://doi.org/10.1007/978-3-031-19067-4_6

3. Due to difference in D_i across workers, the resulting models $\mathbf{w}_{t+\tau}^{(i)}$ will be different
4. The resulting models $\mathbf{w}_{t+\tau}^{(i)}$ are aggregated by the parameter server as per the update rule

$$\mathbf{w}_{t+\tau} = \sum_{i=1}^{m} \frac{\mathbf{w}_{t+\tau}^{(i)}}{m}.$$

The above steps repeat again in the next round, where the workers fetch the aggregated model $\mathbf{w}_{t+\tau}$ from the parameter server.

Thus, suppose all the local models are initialized to the same value $\mathbf{w}_1^{(i)} = \mathbf{w}_1$ for all $i = 1, \ldots, m$, the local-update SGD update rule in the t-th iteration can be written formally as follows:

$$\mathbf{w}_{t+1}^{(i)} = \begin{cases} \frac{1}{m} \sum_{j=1}^{m} \left[\mathbf{w}_t^{(j)} - \eta g(\mathbf{w}_t^{(j)}) \right] & \text{if } t \mod \tau = 0 \\ \mathbf{w}_t^{(i)} - \eta g(\mathbf{w}_t^{(i)}) & \text{otherwise} \end{cases} \quad (6.1)$$

where $g(\mathbf{w}_t^{(i)})$ represents a mini-batch stochastic gradient of the objective function $F(\mathbf{w}_t^{(i)})$ computed using a batch of b datapoints sampled uniformly at random with replacement from worker i's local dataset.

Beyond data-center-based training using the parameter server framework, local-update SGD is especially relevant and well-suited for the emerging framework of federated learning [1–4], where the model is trained using local computation capabilities and data at edge devices such as cell phones or sensors. Since these devices are typically connected to the central (parameter) server via limited bandwidth and high latency wireless communication links, using a communication-efficient algorithm is especially important to ensure fast and scalable distributed training.

Figures 6.1 and 6.2 present two different views of how local updates affect the performance of the algorithm in terms of the runtime and convergence respectively. The pink colored block in Fig. 6.1 represents the time taken by the parameter server to aggregate the

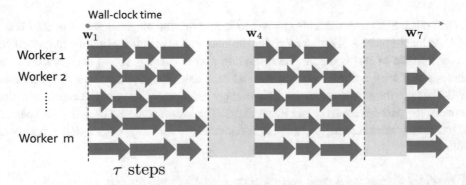

Fig. 6.1 Timeline of the local-update SGD, illustrating how performing more local updates (larger τ) reduces the runtime per iteration. This is because the communication delay per round (represented by the cream colored rectangles) is amortized across the τ iterations within each round

Fig. 6.2 Local-update SGD illustration

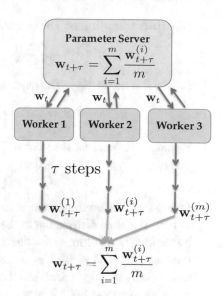

local models $\mathbf{w}_{t+\tau}^{(i)}$ from workers $i = 1, \ldots, m$. The blue arrows at each worker are the τ local updates made to the model $\mathbf{w}_t$ that the workers receive from the parameter server at the beginning of the communication round. By communicating only once in τ iterations, the local-update SGD algorithm amortizes the communication delay across the τ iterations, thus reducing the runtime per iteration. In Sect. 6.1.2 we will formally analyze the runtime per iteration of local SGD to understand how it is affected by τ.

While it has a faster runtime per iteration, local-update SGD has worse error convergence than synchronous distributed SGD because of infrequent consensus between the models at the m workers. This phenomenon is illustrated in Fig. 6.2. As τ increases, the m worker models drift further away from each other because of the differences in their local data partitions. This causes a higher variance in the aggregated model updates, which in turn results in a worse error floor at convergence. In Sect. 6.1.1 we will analyze the error convergence of local-update SGD and quantify how the rate of convergence and the error floor are affected by τ.

6.1.1 Convergence Analysis

Local-update SGD optimizes the same empirical risk objective function $F(h) = \frac{1}{N} \sum_{n=1}^{N} \ell(h(\mathbf{x}_n), y_n)$ that we considered in previous chapters. If decomposed in We make the following assumptions about the objective function $F(\mathbf{w})$ in order to analyze the convergence of local-update SGD.

1. **Lipschitz Smoothness**: The objective function $F(\mathbf{w})$ is differentiable and L-Lipschitz smooth, that is,

$$\|\nabla F(\mathbf{w}) - \nabla F(\mathbf{w}')\| \leq L\|\mathbf{w} - \mathbf{w}'\|$$

2. **Lower Bound on F_{inf}**: The objective function $F(\mathbf{w})$ is bounded below by F_{inf}, that is,

$$F(\mathbf{w}) \geq F_{inf} \text{ for all } \mathbf{w}$$

3. **Unbiased Gradients**: The stochastic gradient $g(\mathbf{w}; \xi)$ is an unbiased estimate of $\nabla F(\mathbf{w})$, that is,

$$\mathbb{E}_\xi[g(\mathbf{w}; \xi)] = \nabla F(\mathbf{w})$$

In order for this assumption to be true, the local datasets $\mathcal{D}_i$ at each of the workers need to have the same distribution as the global dataset $\mathcal{D} = \cup_{i=1}^m \mathcal{D}_i$. This can be achieved by uniformly shuffling and splitting the dataset $\mathcal{D}$ across the workers. However, when local-update SGD is used in the federated learning framework, the local datasets are collected by the edge clients and they inherently have different distributions. Therefore, when local-update SGD is analyzed in the federated learning setting, this assumption is replaced by a local unbiasedness assumption $\mathbb{E}_\xi[g_i(\mathbf{w}; \xi)] = \nabla F_i(\mathbf{w})$, along with an additional bounded heterogeneity assumption on the difference between the gradients across the edge workers.

4. **Bounded Variance**: The stochastic gradient $g(\mathbf{w}; \xi)$ has bounded variance, that is,

$$\text{Var}(g(\mathbf{w}; \xi)) \leq \frac{\sigma^2}{b}$$

$$\mathbb{E}_\xi[\|g(\mathbf{w}; \xi)\|^2] \leq \|\nabla F(\mathbf{w})\|^2 + \frac{\sigma^2}{b}$$

Theorem 6.1 (Convergence of Local-update SGD)   *For a L-smooth function, $\eta L + \eta^2 L^2 \tau(\tau - 1) \leq 1$, and if the starting point is $\mathbf{w}_1$ then after t iterations of local update SGD we have*

$$\mathbb{E}\left[\frac{1}{t}\sum_{k=1}^{t}\|\nabla F(\mathbf{w}_k)\|^2\right] \leq \frac{2\left[F(\mathbf{w}_1) - F_{inf}\right]}{\eta t} + \frac{\eta L \sigma^2}{mb} + \frac{\eta^2 L^2 \sigma^2(\tau - 1)}{b},$$

where $\mathbf{w}_k = \sum_{i=1}^m \mathbf{w}_k^{(i)}/m$ denotes the averaged model at the kth iteration. This is a virtual sequence of iterates that is equal to the global model after every τ iterations when k mod $\tau = 0$.

The version of the local-update SGD analysis presented in Theorem 6.1 above is based on [5], which does not assume strong convexity of the objective function F. That is why, the left hand side is the average of gradients rather than the optimality gap $F(\mathbf{w}) - F^*$, because for a non-convex objective, SGD can get stuck in a stationary point and it is not guaranteed to convergence to the optimal value F^*. Another version of the analysis with the strong convexity assumption is given in [6].

6.1.1.1 Implications of the Convergence Result

For the special case when the communication period $\tau = 1$, this bound reduces the convergence bound of mini-batch SGD for non-convex objective functions. The last term in the error floor grows linearly with τ, corroborating our intuition that the error floor increases with τ due to the increased discrepancy in the model versions at the workers after the local training iterations.

While the error floor does increase if we consider a constant learning rate that is independent of the other parameters, if we set the learning rate as $\eta = \frac{1}{L}\sqrt{\frac{m}{t}}$ and $10m\tau^2 \leq t$, where t is the total number of iterations for which we plan to run the algorithm, then

$$\mathbb{E}\left[\frac{1}{t}\sum_{k=1}^{t}\|\nabla F(\mathbf{w}_k)\|^2\right] \leq \frac{2L\left[F(\mathbf{w}_1) - F_{\text{inf}}\right] + \sigma^2/b}{\sqrt{mt}} + \frac{m(\tau-1)\sigma^2}{tb} \tag{6.2}$$

$$= \mathcal{O}(\frac{1}{\sqrt{mt}}) + \mathcal{O}\left(\frac{m(\tau-1)}{t}\right). \tag{6.3}$$

The second term in (6.3) decays faster with the number of iterations t than the first term. Thus, the overall convergence rate is $\mathcal{O}(\frac{1}{\sqrt{mt}})$. Due to the m in the denominator, we can conclude that local SGD provides a linear speed-up in convergence with respect to the number of workers m. This means that using m workers to perform local updates in parallel results in convergence in m times fewer iterations than the single-node case.

6.1.1.2 Proof of Theorem 6.1

To prove the convergence bound for local-update SGD, let us define the following terms:

- $\mathcal{G}_k = \frac{1}{m}\sum_{i=1}^{m} g(\mathbf{w}_k^{(i)})$, the average of stochastic gradients at the workers at iteration k, and
- $\mathcal{H}_k = \frac{1}{m}\sum_{i=1}^{m} \nabla F(\mathbf{w}_k^{(i)})$, the average of the full gradients at the workers at iteration k.

For the ease of writing, we define the notation $\mathbb{E}_k$ to denote the conditional expectation $\mathbb{E}_{\Xi_K|\mathbf{z}_k}$, where Ξ_k is the set $\{\xi_k^{(1)}, \ldots, \xi_k^{(m)}\}$ of mini-batches at m workers in iteration k. By the unbiasedness of the gradients, observe that $\mathbb{E}_k[\mathcal{G}_k] = \mathcal{H}_k$.

Using the definitions of $\mathcal{G}_k$ and $\mathcal{H}_k$, we can show that

$$\mathbb{E}_k \left[\| \mathcal{G}_k - \mathcal{H}_k \|^2 \right] \tag{6.4}$$

$$= \mathbb{E}_k \left\| \frac{1}{m} \sum_{i=1}^m \left[g(\mathbf{w}_k^{(i)}) - \nabla F(\mathbf{w}_k^{(i)}) \right] \right\|^2 \tag{6.5}$$

$$= \frac{1}{m^2} \mathbb{E}_k \left[\sum_{i=1}^m \left\| g(\mathbf{w}_k^{(i)}) - \nabla F(\mathbf{w}_k^{(i)}) \right\|^2 \right] +$$

$$\sum_{j \neq l}^m \mathbb{E}_k \left[\langle g(\mathbf{w}_k^{(j)}) - \nabla F(\mathbf{w}_k^{(j)}) g(\mathbf{w}_k^{(l)}) - \nabla F(\mathbf{w}_k^{(l)}) \rangle \right] \tag{6.6}$$

$$\leq \frac{\sigma^2}{bm}, \tag{6.7}$$

where Eq. (6.7) is due to $\{\xi_k^{(i)}\}$ are independent random variables. By directly applying the unbiased gradient and bounded variance assumptions to (6.6), one can observe that all cross terms are zero.

Using the result in (6.7), and the fact that $\mathbb{E}_k[\mathcal{G}_k] = \mathcal{H}_k$, we can also show that:

$$\mathbb{E}_k \left[\| \mathcal{G}_k \|^2 \right] = \mathbb{E}_k \left[\| \mathcal{G}_k - \mathbb{E}_k[\mathcal{G}_k] \|^2 + \| \mathbb{E}_k[\mathcal{G}_k] \|^2 \right] \tag{6.8}$$

$$= \mathbb{E}_k \left[\| \mathcal{G}_k - \mathcal{H}_k \|^2 + \| \mathcal{H}_k \|^2 \right] \tag{6.9}$$

$$\leq \frac{\sigma^2}{bm} + \| \mathcal{H}_k \|^2 \tag{6.10}$$

$$= \frac{\sigma^2}{bm} + \mathbb{E} \left[\| \frac{1}{m} \sum_{i=1}^m \nabla F(\mathbf{w}_k^{(i)}) \|^2 \right] \tag{6.11}$$

$$\leq \frac{1}{m} \sum_{i=1}^m \| \nabla F(\mathbf{w}_k^{(i)}) \|^2 + \frac{\sigma^2}{bm}, \tag{6.12}$$

where the last inequality follows from the convexity of vector norm and Jensen's inequality. Another property of $\mathcal{G}_k$ required for the convergence analysis is that

$$\mathbb{E} \left[\langle \nabla F(\mathbf{w}_k), \mathcal{G}_k \rangle \right] = \mathbb{E} \left[\langle \nabla F(\mathbf{w}_k), \frac{1}{m} \sum_{i=1}^m g(\mathbf{w}_k^{(i)}) \rangle \right] \tag{6.13}$$

$$= \frac{1}{m} \sum_{i=1}^m \langle \nabla F(\mathbf{w}_k), \nabla F(\mathbf{w}_k^{(i)}) \rangle \tag{6.14}$$

$$= \frac{1}{2m} \sum_{i=1}^m \left[\| \nabla F(\mathbf{w}_k) \|^2 + \| \nabla F(\mathbf{w}_k^{(i)}) \|^2 \right.$$

$$\left. - \| \nabla F(\mathbf{w}_k) - \nabla F(\mathbf{w}_k^{(i)}) \|^2 \right] \tag{6.15}$$

$$= \frac{1}{2} \|\nabla F(\mathbf{w}_k)\|^2 + \frac{1}{2m} \sum_{i=1}^{m} \|\nabla F(\mathbf{w}_k^{(i)})\|^2$$

$$- \frac{1}{2m} \sum_{i=1}^{m} \|\nabla F(\mathbf{w}_k) - \nabla F(\mathbf{w}_k^{(i)})\|^2, \qquad (6.16)$$

where we use the fact that $2\mathbf{w}^T \mathbf{w}' = \|\mathbf{w}\|^2 + \|\mathbf{w}'\|^2 - \|\mathbf{w} - \mathbf{w}'\|^2$.

Now we will use the bounds derived in (6.12) and (6.16) to analyze the difference between the values of two successive values $\mathbb{E}_k[F(\mathbf{w}_{k+1})]$ and $F(\mathbf{w}_k)$ of the objective function where $\mathbb{E}_k$ denotes the conditional expectation given all the mini-batches processed until iteration k. We start with the Lipschitz smoothness assumption, which implies the following.

$$\mathbb{E}\left[F(\mathbf{w}_{k+1})\right] - F(\mathbf{w}_k)$$

$$\leq -\eta \mathbb{E}[\langle \nabla F(\mathbf{w}_k), \mathcal{G}_k \rangle] + \frac{\eta^2 L}{2} \mathbb{E}[\|\mathcal{G}_k\|^2] \qquad (6.17)$$

$$\leq -\frac{\eta}{2} \|\nabla F(\mathbf{w}_k)\|^2 - \frac{\eta}{2m} \sum_{i=1}^{m} \|\nabla F(\mathbf{w}_k^{(i)})\|^2 + \frac{\eta}{2m} \sum_{i=1}^{m} \|\nabla F(\mathbf{w}_k) - \nabla F(\mathbf{w}_k^{(i)})\|^2$$

$$+ \frac{\eta^2 L \sigma^2}{2bm} + \frac{\eta^2 L}{2m} \sum_{i=1}^{m} \|\nabla F(\mathbf{w}_k^{(i)})\|^2 \qquad (6.18)$$

$$\leq -\frac{\eta}{2} \|\nabla F(\mathbf{w}_k)\|^2 - \frac{\eta(1 - \eta L)}{2m} \sum_{i=1}^{m} \|\nabla F(\mathbf{w}_k^{(i)})\|^2 + \frac{\eta L^2}{2m} \sum_{i=1}^{m} \|\mathbf{w}_k - \mathbf{w}_k^{(i)}\|^2 + \frac{\eta^2 L \sigma^2}{2bm}, \qquad (6.19)$$

where (6.18) uses the result in (6.16), and (6.19) follows from the Lipschitz smoothness of F. Rearranging the terms to bring $\|\nabla F(\mathbf{w}_k)\|^2$ to the left hand side, we have

$$\|\nabla F(\mathbf{w}_k)\|^2 \leq \frac{2(F(\mathbf{w}_k) - \mathbb{E}\left[F(\mathbf{w}_{k+1})\right])}{\eta} - \frac{(1 - \eta L)}{m} \sum_{i=1}^{m} \|\nabla F(\mathbf{w}_k^{(i)})\|^2 +$$

$$\frac{L^2}{m} \sum_{i=1}^{m} \|\mathbf{w}_k - \mathbf{w}_k^{(i)}\|^2 + \frac{\eta L \sigma^2}{bm} \qquad (6.20)$$

$$\qquad (6.21)$$

Now taking total expectation and averaging over t iterations,

$$\frac{1}{t}\sum_{k=1}^{t}\mathbb{E}[\|\nabla F(\mathbf{w}_k)\|^2] \leq \underbrace{\frac{2(F(\mathbf{w}_1) - F_{inf})}{t\eta} + \frac{\eta L\sigma^2}{bm} +}_{\text{synchronous SGD error}}$$

$$\underbrace{\frac{L^2}{mt}\sum_{k=1}^{t}\sum_{i=1}^{m}\mathbb{E}[\|\mathbf{w}_k - \mathbf{w}_k^{(i)}\|^2] - \frac{(1-\eta L)}{mt}\sum_{k=1}^{t}\sum_{i=1}^{m}\mathbb{E}[\|\nabla F(\mathbf{w}_k^{(i)})\|^2]}_{\text{additional error}} \qquad (6.22)$$

The additional error term arises because we only average after every τ local iterations, which causes a discrepancy between the averaged model $\mathbf{w}_k$ and the local model $\mathbf{w}_k^{(i)}$ at the i-th worker. The Cooperative SGD paper [5] gives an upper bound on this additional error term and shows that

$$\frac{L^2}{mt}\sum_{k=1}^{t}\sum_{i=1}^{m}\mathbb{E}[\|\mathbf{w}_k - \mathbf{w}_k^{(i)}\|^2] - \frac{(1-\eta L)}{mt}\sum_{k=1}^{t}\sum_{i=1}^{m}\mathbb{E}[\|\nabla F(\mathbf{w}_k^{(i)})\|^2] \leq \frac{\eta^2 L^2\sigma^2(\tau-1)}{b}$$

$$(6.23)$$

when the learning rate is small enough. The proof of (6.23) is as follows.

To write the above expression more compactly, define matrices $\mathbf{W}_k, \mathbf{G}_k \in \mathbb{R}^{d\times m}$ that are concatenate the models and the gradients at the m workers:

$$\mathbf{W}_k = [\mathbf{w}_k^{(1)}, \ldots, \mathbf{w}_k^{(m)}] \qquad (6.24)$$

$$\mathbf{G}_k = [g(\mathbf{w}_k^{(1)}), \ldots, g(\mathbf{w}_k^{(m)})] \qquad (6.25)$$

We also define matrix $\mathbf{J} = \mathbf{1}\mathbf{1}^\top/(\mathbf{1}^\top\mathbf{1})$ where $\mathbf{1}$ is the all-ones column vector. Thus, every element of the matrix $\mathbf{J}$ is equal to $1/m$. As usual $\mathbf{I}$ denotes the identity matrix of size $m \times m$, where m is the number of workers. Moreover, for a matrix $\mathbf{A}$, $\|\mathbf{A}\|_F^2$ denotes its Frobenius norm, which is the sum of the squares of each of the elements of the matrix. And the operator norm $\|\mathbf{A}\|_{op}$ of a matrix is defined as $\max_{\|\mathbf{x}\|=0}\|\mathbf{A}\mathbf{x}\| = \sqrt{\lambda_{max}(\mathbf{A}^\top\mathbf{A})}$.

Let us analyze the term $\sum_{i=1}^{m}\mathbb{E}[\|\mathbf{w}_k - \mathbf{w}_k^{(i)}\|^2]$ for different values of the iteration index k, by representing $k = r\tau + a$, where $r \geq 0$ is the communication round index, and $1 \leq a \leq \tau$ is the index of a local step within that communication round.

$$\sum_{i=1}^{m}\mathbb{E}[\|\mathbf{w}_{r\tau+a} - \mathbf{w}_{r\tau+a}^{(i)}\|^2] \qquad (6.26)$$

$$= \mathbb{E}[\|\mathbf{W}_{r\tau+a}(\mathbf{I} - \mathbf{J})\|^2] \qquad (6.27)$$

$$= \mathbb{E}[\|\mathbf{W}_{r\tau+1}(\mathbf{I} - \mathbf{J}) - \eta\sum_{l=1}^{a-1}\mathbf{G}_{r\tau+l}(\mathbf{I} - \mathbf{J})\|^2] \qquad (6.28)$$

$$= \mathbb{E}[\|-\eta\sum_{l=1}^{a-1}\mathbf{G}_{r\tau+l}(\mathbf{I} - \mathbf{J})\|^2] \qquad (6.29)$$

$$\leq \eta^2 \mathbb{E}[\|\sum_{l=1}^{a-1} \mathbf{G}_{r\tau+l}\|^2] \tag{6.30}$$

$$= \eta^2 \sum_{i=1}^{m} \mathbb{E}[\|\sum_{l=1}^{a-1} g(\mathbf{w}_{r\tau+l}^{(i)})\|^2] \tag{6.31}$$

$$\leq 2\eta^2 \sum_{i=1}^{m} \mathbb{E}[\|\sum_{l=1}^{a-1} (g(\mathbf{w}_{r\tau+l}^{(i)}) - \nabla F(\mathbf{w}_{r\tau+l}^{(i)})\|^2] + 2\eta^2 \sum_{i=1}^{m} \mathbb{E}[\|\sum_{l=1}^{a-1} \nabla F(\mathbf{w}_{r\tau+l}^{(i)})\|^2] \tag{6.32}$$

$$\leq 2\eta^2 m(a-1)\frac{\sigma^2}{b} + 2\eta^2 \sum_{i=1}^{m} \mathbb{E}[\|\sum_{l=1}^{a-1} \nabla F(\mathbf{w}_{r\tau+l}^{(i)})\|^2] \tag{6.33}$$

$$\leq 2\eta^2 m(a-1)\frac{\sigma^2}{b} + 2\eta^2(a-1) \sum_{i=1}^{m} \sum_{l=1}^{a-1} \mathbb{E}[\|\nabla F(\mathbf{w}_{r\tau+l}^{(i)})\|^2] \tag{6.34}$$

where in (6.30) we use the fact that the operator norm of $(\mathbf{I} - \mathbf{J})$ is less than or equal to 1. To get (6.32) we use the fact that $(a + b)^2 \leq 2a^2 + 2b^2$. In (6.33) we use the bounded gradient variance assumption. And in (6.38) we use the inequality $(a_1 + a_2 + \cdots + a_m)^2 \leq m(a_1^2 + \cdots + a_m^2)$.

Summing over all a and r and multiplying by L^2/mt we have:

$$\frac{L^2}{mt} \sum_{k=1}^{t} \sum_{i=1}^{m} \mathbb{E}[\|\mathbf{w}_k - \mathbf{w}_k^{(i)}\|^2] \tag{6.35}$$

$$\leq \frac{L^2}{mt} \sum_{r=0}^{\lfloor t/\tau \rfloor} \sum_{a=1}^{\min(\tau,k-r\tau)} \sum_{i=1}^{m} \mathbb{E}[\|\mathbf{w}_{r\tau+a} - \mathbf{w}_{r\tau+a}^{(i)}\|^2] \tag{6.36}$$

$$\leq \frac{2\eta^2 L^2}{mt} \sum_{r=0}^{\lfloor t/\tau \rfloor} \sum_{a=1}^{\min(\tau,k-r\tau)} m(a-1)\frac{\sigma^2}{b} +$$

$$\frac{2\eta^2 L^2}{mt} \sum_{r=0}^{\lfloor t/\tau \rfloor} \sum_{a=1}^{\min(\tau,k-r\tau)} (a-1) \sum_{i=1}^{m} \sum_{l=1}^{a-1} \mathbb{E}[\|\nabla F(\mathbf{w}_{r\tau+l}^{(i)})\|^2] \tag{6.37}$$

$$\leq 2\frac{\eta^2 L^2}{t} \frac{t}{\tau} \frac{\tau(\tau-1)\sigma^2}{2b} + \frac{\eta^2 L^2}{mt} \sum_{i=1}^{m} \sum_{k=1}^{t} \tau(\tau-1)\mathbb{E}[\|\nabla F(\mathbf{w}_k^{(i)})\|^2] \tag{6.38}$$

$$= \frac{\eta^2 L^2(\tau-1)\sigma^2}{b} + \frac{\eta^2 L^2\tau(\tau-1)}{mt} \sum_{i=1}^{m} \sum_{k=1}^{t} \mathbb{E}[\|\nabla F(\mathbf{w}_k^{(i)})\|^2]. \tag{6.39}$$

Substituting this in (6.22) we have:

$$\frac{1}{t} \sum_{k=1}^{t} \mathbb{E}[\|\nabla F(\mathbf{w}_k)\|^2] \leq \frac{2(F(\mathbf{w}_1) - F_{inf})}{t\eta} + \frac{\eta L\sigma^2}{bm} +$$

$$\frac{\eta^2 L^2(\tau - 1)\sigma^2}{b} - \frac{(1 - \eta L - \eta^2 L^2 \tau(\tau - 1))}{mt} \sum_{k=1}^{t} \sum_{i=1}^{m} \mathbb{E}[\|\nabla F(\mathbf{w}_k^{(i)})\|^2] \quad (6.40)$$

$$\leq \frac{2(F(\mathbf{w}_1) - F_{inf})}{t\eta} + \frac{\eta L \sigma^2}{bm} + \frac{\eta^2 L^2(\tau - 1)\sigma^2}{b} \quad (6.41)$$

where we can drop the last term in (6.40) if the learning rate η is small enough such that $\eta L + \eta^2 L^2 \tau(\tau - 1) < 1$. Thus, we get the upper bound on the additional term as given in (6.23). Finally, by combining (6.22) and (6.23) we have proved Theorem 6.1.

6.1.2   Runtime Analysis

In Sect. 6.1.1 we analyzed the error versus iterations convergence of local-update SGD and showed that the error versus iterations covergence degrades with τ because the discrepancy between the local models results in a higher error floor. In the section, we consider the other side of this story, and analyze the effect of τ on the runtime per iteration.

6.1.2.1 Delay Model

At the beginning of a communication round, all the m workers receive the current version of the model $\mathbf{w}$ and commence τ local updates each. We use the random variable $Y \sim F_Y$ denotes the taken by a worker to perform one local update, which includes the time taken to compute one mini-batch gradient and update the model. Assume that these local update times are i.i.d. across workers and the τ local updates at each worker. Thus, if Y_{ij} is time taken to finish the j-th local update in the round at the i-th worker. Thus, Y_{ij} for all $i = 1, \ldots, m$ and $j = 1, 2, \ldots, \tau$ are i.i.d. realizations of the probability distribution F_Y.

After all the workers finish their local updates, their models are sent to the parameter server, which averages them and sends back the workers. We refer to the time window from the instant when the latest worker finishes and sends its local updates, until all the workers receive the new model version as the *communication delay*. It is modeled by a random variable $D \sim F_D$, which is assumed to be i.i.d. across communication rounds.

6.1.2.2 Total Local Computation Time per Round

Using the delay model described above, let us compute the total local computation time T_{comp} of a round, which is defined as the time from the start of the local updates at the m workers, until the last worker finishes its τ local updates.

$$T_{comp} = \max_{i=1,\ldots,m} \left(Y_{i,1} + Y_{i,2} + \cdots + Y_{i,\tau} \right). \quad (6.42)$$

In order to rewrite T_{comp} more compactly and analyze how its probability distribution depends on F_Y, we define a random variable Y_i' which denotes that average of the τ local update times are worker i, that is,

$$Y_i' = \frac{\left(Y_{i,1} + Y_{i,2} + \cdots + Y_{i,\tau}\right)}{m} \tag{6.43}$$

Since $Y_{i,j}$s are i.i.d., the average local update time random variables Y_i' are also i.i.d. across workers. To understand the difference between the probability distributions F_Y and $F_{Y'}$, let us compare their expected values and variances:

$$\mathbb{E}[Y'] = \mathbb{E}[Y] \tag{6.44}$$

$$\mathrm{Var}\left[Y'\right] = \frac{\mathrm{Var}\,[Y]}{\tau} \tag{6.45}$$

Thus, the random variables Y' has the same mean but τ times lower variance than Y.

The expected computation time T_{comp} can thus be expressed in terms Y' as follows:

$$\mathbb{E}\left[T_{comp}\right] = \tau \mathbb{E}\left[Y_{m:m}'\right] \tag{6.46}$$

where $Y_{m:m}' = \max(Y_1', \ldots Y_m')$, the maximum order statistic of the average local update times $Y_1', \ldots Y_m'$ at the m workers.

6.1.2.3 Expected Runtime Per Iteration

The time taken to complete one communication round is the total local computation time T_{comp} plus the communication delay D. Since τ iterations are completed in each round, the expected runtime per iteration is:

$$\mathbb{E}[T_{local}] = \frac{\mathbb{E}\left[T_{comp}\right] + \mathbb{E}\,[D]}{\tau} \tag{6.47}$$

$$= \mathbb{E}[Y_{m:m}'] + \frac{\mathbb{E}[D]}{\tau} \tag{6.48}$$

Let us compare this with the runtime of synchronous SGD

$$\mathbb{E}[T_{sync}] = \mathbb{E}[Y_{m:m}] + \mathbb{E}[D] \tag{6.49}$$

By comparing (6.48) and (6.49), we observe that performing multiple local updates at each worker and averaging the models periodically reduces the expected runtime in two ways.

Firstly, the communication delay $\mathbb{E}\,[D]$ gets divided by τ because it gets amortized across τ iterations completed in each communication round. In contrast, synchronous SGD has to incur the communication delay after every iteration. This reduction in communication delay is a game-changer in bandwidth-limited settings such as federated learning, where the worker

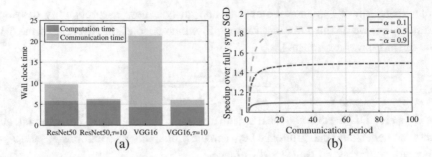

Fig. 6.3 **a** Wallclock time to finish 100 iterations in a cluster with 4 worker nodes. To achieve the same computation/computation ratio, VGG-16 requires a larger communication period τ than ResNet-50. **b** The speed-up offered by using periodic-averaging SGD increases with τ (the communication period) and with the communication/computation delay ratio $\alpha = D/Y$, where D is the all-node broadcast delay and Y is the time taken for each local update at a worker

nodes are wirelessly connected mobile devices. Also, the relative values of the means of Y and D depend on the size of the model being trained. Suppose Y and D are constants, and their ratio, the communication to computation time ratio is $\alpha = D/Y$. The value of α depends on the size of the model being trained, the available communication bandwidth and other system parameters. For experiments run on a university computing cluster, we observed that the value of α for the VGG16 model is more than that of ResNet16 because VGG16 is a larger neural network, as shown in Fig. 6.3a. The speed-up of local-update SGD over synchronous SGD is given by

$$\frac{\mathbb{E}[T_{sync}]}{\mathbb{E}[T_{local}]} = \frac{Y + D}{Y + D/\tau} = \frac{1 + \alpha}{1 + \alpha/\tau} \tag{6.50}$$

As α increases, local-update SGD gives an even larger speed-up over synchronous SGD. Figure 6.3b shows the speed-up for different values of α and τ. When D is comparable with Y ($\alpha = 0.9$), local-update SGD can be almost twice as fast as fully synchronous SGD.

Local SGD also reduces the runtime per iteration in another, more subtle way. When we perform multiple local updates, the variations across those updates are *evened out* as τ becomes larger, thus providing a straggler mitigation effect. This can be formally observed from the fact that the value of the maximum order statistic $\mathbb{E}[Y'_{m:m}]$ is less than $\mathbb{E}[Y_{m:m}]$ because Y' has lower variance than Y. This reduction is more drastic when the distribution of Y has a heavier tail and/or for a larger number of workers m. In Fig. 6.4, we demonstrate this phenomenon for exponentially distributed Y's. Local-update SGD not only gives an almost two-fold reduction in the expected runtime per iteration (shown by the dotted line but also reduces the tail of the runtime distribution. In practice, the assumption that the Y's are i.i.d. across local updates at a worker may not be true because a worker that slows down may remain slow for several consecutive local iterations. As a result, the reduction in the expected runtime may not be as drastic as illustrated in Fig. 6.4. However we still expect local-update SGD to provide a straggler mitigation benefit in addition to the communication delay amortization.

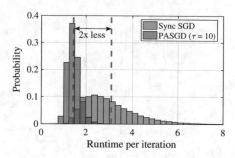

Fig. 6.4 Comparison of the probability distributions of the runtime per iteration of local update SGD ($\tau = 10$), also called periodic averaging SGD (PASGD), with synchronous SGD for exponentially distributed local update times $Y \sim \text{Exp}(1)$ and constant communication delay $D = 1$ for a system of $m = 16$ workers. Local-update SGD not only reduces the mean (shown by the dotted line) but also reduces the tail latency

6.1.3   Adaptive Communication

The convergence and runtime analyses of local-update SGD in Sects. 6.1.1 and 6.1.2 show that there is an inherent error-runtime trade-off in terms of the number of local updates τ. Experimental results match this insight, as shown in Fig. 6.5. To obtain these results, we train the VGG-16 [7] neural network on the CIFAR-10 dataset [8] with 8 worker nodes in a university computing cluster. We start with a constant learning rate and then reduce it by a factor of 10 after 100 and 150 epochs (traversals of the entire dataset). If we look at the training loss versus epoch plots in Fig. 6.5 we see that increasing τ results in a higher error floor, as suggested by our convergence analysis in Sect. 6.1.1. However, if we account for the runtime per iteration and plot the training loss versus wallclock time, we observe that larger τ provides a drastic runtime reduction. These error versus walllclock time plots suggest a natural strategy to enjoy fast runtime per iteration as well as a small error floor at convergence—*starting with a larger τ and gradually reducing it during the training process.*

6.1.3.1 Derivation of the Adacomm Strategy
Next, let us develop this strategy called Adacomm, which determines the best τ at each wallclock time T. Using this strategy, we will switch from one learning curve to another, as illustrated in Fig. 6.6. First, we need to determine the error after T iterations. From the convergence analysis in Sect. 6.1.1 we know that the error after t iterations is bounded by

$$\mathbb{E}\left[\frac{1}{t}\sum_{k=1}^{t}\|\nabla F(\mathbf{w}_k)\|^2\right] \leq \frac{2\left[F(\mathbf{w}_1) - F_{\inf}\right]}{\eta t} + \frac{\eta L \sigma^2}{mb} + \frac{\eta^2 L^2 \sigma^2 (\tau - 1)}{b} \qquad (6.51)$$

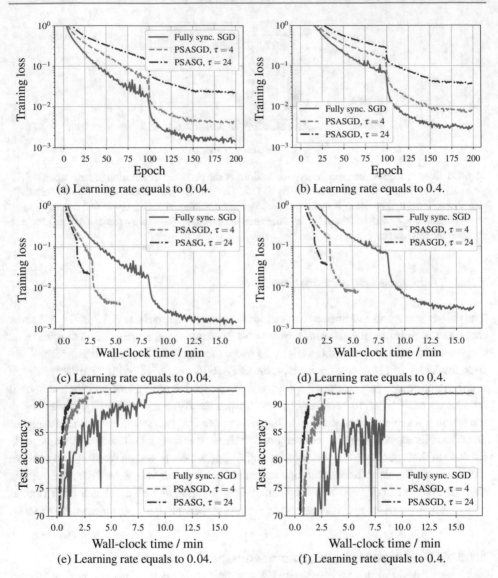

Fig. 6.5 Illustration of error-convergence and communication-efficiency trade-off in local-update SGD. We train a VGG-16 on CIFAR-10 with 8 worker nodes. Each line was trained for 200 epochs (traversals of the training dataset) and the learning rate is decayed by 10 at epoch 100, 150. After the same number of epochs, a larger communication period leads to higher training loss but costs much less wall clock time

Suppose that the local update time Y and the communication delay D are constants, so that the runtime per iteration of local-update SGD is $T_{local} = Y + D/\tau$. Substituting this in the error bound, we can show that the error at time T is bounded by

$$\text{Error at time } T \leq \frac{2\left[F(\mathbf{w}_1) - F_{inf}\right]}{\eta T}\left(Y + \frac{D}{\tau}\right) + \frac{\eta L \sigma^2}{mb} + \frac{\eta^2 L^2 \sigma^2 (\tau - 1)}{b} \quad (6.52)$$

We derive a heuristic schedule for τ by minimizing this upper bound with respect to τ. Setting the derivative of the right hand side with respect to τ to 0, we get

$$-\frac{2\left[F(\mathbf{w}_1) - F_{inf}\right]}{\eta T}\left(\frac{D}{\tau^2}\right) + \frac{\eta^2 L^2 \sigma^2}{b} = 0 \quad (6.53)$$

$$\tau^* = \sqrt{\frac{2b\left[F(\mathbf{w}_1) - F_{inf}\right] D}{\eta^3 L^2 \sigma^2 T}} \quad (6.54)$$

6.1.3.2 Modifying the Schedule to Accommodate Practical Constraints

Can we use directly use the τ schedule given by (6.54) in practice? Unfortunately, not, due to several parameters ($F(\mathbf{w}_1) - F_{inf}$), L, σ^2 which are properties of the objective function and its stochastic gradients are not known in practice. Also, it would be impossible to change τ at every time instant T. We accommodate these practical constraints by first dividing time into intervals of size T_0 and finding best τ for each interval Fig. 6.6. In the first interval, the τ_0^* is

$$\tau_0^* = \sqrt{\frac{2b\left[F(\mathbf{w}_{T=0}) - F_{inf}\right] D}{\eta^3 L^2 \sigma^2 T_0}} \quad (6.55)$$

Then, we find τ_l by considering that the workers start at the initial point $F(\mathbf{w}_{T=lT_0})$:

$$\tau_l^* = \sqrt{\frac{2b\left[F(\mathbf{w}_{T=lT_0}) - F_{inf}\right] D}{\eta^3 L^2 \sigma^2 T_0}} \quad (6.56)$$

Now taking the ratio of τ_l^* and τ_0^* we get:

$$\tau_l^* = \sqrt{\frac{\left[F(\mathbf{w}_{T=lT_0}) - F_{inf}\right]}{\left[F(\mathbf{w}_{T=0}) - F_{inf}\right]}}\tau_0^* \sim \sqrt{\frac{F(\mathbf{w}_{T=lT_0})}{F(\mathbf{w}_{T=0})}}\tau_0^* \quad (6.57)$$

where we assume that $F_{inf} = 0$. Observe that the unknown terms L and σ^2 cancel out. The only remaining unknown variable is initial τ_o^*, which we can choose using grid search over several possible values by treating it as a hyperparameter.

Fig. 6.6 Illustration of adaptive
communication strategies. The
dashed line denotes the
learning curve with adaptive
communication

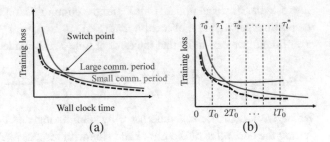

(a)   (b)

6.1.3.3 Incorporating a Variable Learning Rate Schedule

When deriving (6.57), we assumed that the learning rate η is constant throughout training. However, in practice, the learning rate is generally decayed during training according to a variable learning rate schedule? We can incorporate such a learning rate schedule into our Adacomm strategy. Consider that the learning rate in the l-th size T_0 time interval is η_l. Now take the ratio of τ_l^* and τ_0^* to get

$$
\tau_l^* = \sqrt{\frac{\eta_0^3}{\eta_l^3} \frac{\left[F(\mathbf{w}_{T=lT_0}) - F_{\text{inf}}\right]}{\left[F(\mathbf{w}_{T=0}) - F_{\text{inf}}\right]}} \tau_0^* \sim \sqrt{\frac{\eta_0^3}{\eta_l^3} \frac{F(\mathbf{w}_{T=lT_0})}{F(\mathbf{w}_{T=0})}} \tau_0^* \tag{6.58}
$$

where τ_0^* is found via brute-force grid search. Observe that when the learning rate is fixed, τ reduces over time. However, for a variable learning rate, if the learning rate decays at some time instant, the corresponding τ increases. This is because, when each step is smaller we can afford to take more local steps without causing too much discrepancy between the local models $\mathbf{w}_k^{(i)}$ across workers $i = 1, \ldots, m$.

Figure 6.7 presents the results for VGG-16 for both fixed and variable learning rates. A large communication period τ initially results in a rapid drop in the error, but the error finally converges to higher floor. By adapting τ, the proposed ADACOMM scheme strikes the best error-runtime trade-off in all settings. In Fig. 6.7, while fully synchronous SGD takes 33.5 min to reach 4×10^{-3} training loss, ADACOMM costs 15.5 minutes achieving more than $2\times$ speedup. Similarly, in Fig. 6.7, ADACOMM takes 11.5 minutes to reach 4.5×10^{-2} training loss achieving $3.3\times$ speedup over fully synchronous SGD (38.0 min).

However, for ResNet-50, the communication overhead is no longer the bottleneck. For fixed communication period, the negative effect of performing local updates becomes more obvious and cancels the benefit of low communication delay. It is not surprising to see fully synchronous SGD is nearly the best one in the error-runtime plot among all fixed-τ methods. Even in this extreme case, adaptive communication can still have a competitive performance. When combined with learning rate decay, the adaptive scheme is about 1.3 times faster than fully synchronous SGD, 15.0 versus 21.5 minutes to achieve 3×10^{-2} training loss.

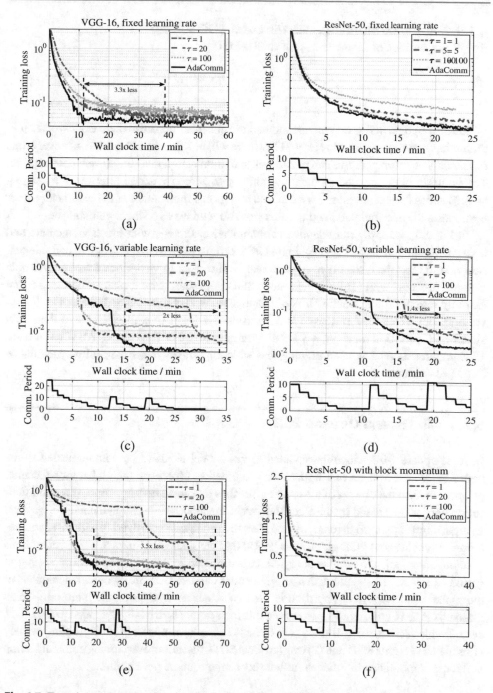

Fig. 6.7 Experimental results showing the performance of Adacomm for the VGG16 (first column) and ResNet-50 (second column) neural networks trained on the CIFAR-10 dataset

6.1.3.4 Applying Momentum in the Local-SGD Setting

Recall the addition of momentum γ in vanilla SGD

$$\mathbf{y}_t = \gamma \mathbf{y}_{t-1} + \eta \nabla F(\mathbf{w}_t)$$
$$\mathbf{w}_{t+1} = \mathbf{w}_t - \mathbf{y}_t$$

In local update SGD, we can use momentum for local updates where each worker will maintain its own momentum buffer $\mathbf{y}_t^{(i)}$. But there will be a dramatic change in $\mathbf{w}_t^{(i)}$ every time the models are averaged; this can create spikes in momentum that can prevent convergence. The solution is to reset the momentum buffer $\mathbf{y}$ to zero at workers after every averaging step. Besides local momentum, we can also add global momentum at the parameter server by treating the accumulated local updates from all workers as a single gradient step.

In Fig. 6.7, we apply our adaptive communication strategy with block momentum and observe significant performance gain on CIFAR10. In particular, the adaptive communication scheme has the fastest convergence rate with respect to wall-clock time in the whole training process. While fully synchronous SGD gets stuck with a plateau before the first learning rate decay, the training loss of adaptive method continuously decreases until converging. For VGG-16, ADACOMM is $3.5\times$ faster (in terms of wall-clock time) than fully synchronous SGD in reaching a 3×10^{-3} training loss. For ResNet-50, ADACOMM takes 15.8 minutes to get 2×10^{-2} training loss which is 2 times faster than fully synchronous SGD (32.6 minutes).

6.2   Elastic and Overlap SGD

In local-update SGD, since the updated global model needs to be communicated to the nodes before the next set of τ updates can commence. Moreover, the global model cannot be updated until the slowest of the m nodes finishes its τ local updates. Since this communication barrier is imposed by the algorithm rather than the systems implementation, we need an algorithmic approach to remove it and allow communication to overlap with local computation. Asynchronous SGD algorithms [9–15] that we studied in Chap. 5, use asynchronous gradient aggregation to remove the synchronization barrier. However, asynchronous aggregation causes model staleness, that is, slow nodes can have arbitrarily outdated versions of the global model. Another approach is to allow faster nodes to compute more mini-batch gradients [15, 16] per iteration. However, this idea cannot be trivially extended to the $\tau > 1$ case. In this section, we will study two variants of the local-update SGD algorithm, namely (1) elastic averaging SGD and (2) overlap local SGD, that allow overlapping communication and local computation in order to further reduce the runtime per iteration.

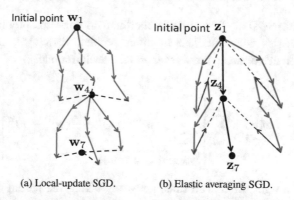

(a) Local-update SGD.   (b) Elastic averaging SGD.

Fig. 6.8 Illustration of local-update SGD and EASGD in the model parameter space. Blue and black arrows denote local update SGD iterations and the update of auxiliary variables, respectively. Red arrows represent averaging local models with each other or with the auxiliary variable. In this toy example, the number of local updates τ is set to 3

6.2.1 Elastic Averaging SGD

In local-update SGD, at averaging step which happens every τ iterations, all the m worker models are reset to the updated model $\mathbf{w}_t = \sum_{i=1}^{m} \mathbf{w}_t^{(i)}/m$. Instead, elastic averaging SGD [17] allows some slack between the worker models. At each averaging step, the parameter server pulls the worker models towards it, and the workers collectively pull the parameter server's model in the opposite direction, as illustrated in Fig. 6.8b. This averaging is analogous to having springs attached between the parameter server and each worker, and these springs exerting elastic force in both directions stopping the parameter server model and each worker model from straying too far away from each other. The elasticity of the spring is represented by the parameter α.

To implement this idea, elastic averaging SGD defines an anchor model $\mathbf{z}$, which is maintained by the parameter server. The update rules for the worker and anchor models are:

$$
\mathbf{w}_{k+1}^{(i)} = \begin{cases} \mathbf{w}_k^{(i)} - \eta g_i(\mathbf{w}_k^{(i)}) - \alpha(\mathbf{w}_k^{(i)} - \mathbf{z}_k), & k \mod \tau = 0 \\ \mathbf{w}_k^{(i)} - \eta g_i(\mathbf{w}_k^{(i)}), & \text{otherwise} \end{cases}, \tag{6.59}
$$

$$
\mathbf{z}_{k+1} = \begin{cases} (1 - m\alpha)\mathbf{z}_k + m\alpha \left(\frac{1}{m} \sum_{i=1}^{m} \mathbf{w}_k^{(i)} \right) & k \mod \tau = 0 \\ \mathbf{z}_k, & \text{otherwise} \end{cases} \tag{6.60}
$$

From (6.60), we observe that the auxiliary variable $\mathbf{z}_k$ can be considered as a moving average of the averaged model $\frac{1}{m} \sum_{i=1}^{m} \mathbf{w}_k^{(i)}$. A larger value of the parameter α forces more consensus between the locally trained models and improves stability, but it may reduce the convergence speed.

Elastic averaging SGD in effect solves a modified optimization problem, instead of the standard empirical risk objective. This modified problem includes a proximal term that penalizes the difference between the worker and the anchor models.

$$
\min_{\mathbf{z}, \mathbf{w}^{(1)}, \dots, \mathbf{w}^{(m)}} \sum_{i=1}^{m} \left[\underbrace{\sum_{n=1}^{N_i} f(\mathbf{w}^{(i)}; \xi_{i,n})}_{\text{Worker } i\text{'s local objective}} + \frac{\rho}{2} \|\mathbf{w}^{(i)} - \mathbf{z}\|^2 \right] \tag{6.61}
$$

where $\rho = \alpha/\eta$. We get the update rules above by taking the partial derivatives of this objective function with respect to $\mathbf{w}^{(i)}$ and $\mathbf{z}$ respectively. Ideas similar to EASGD have been proposed in the context of federated learning, for example FedProx [18], which seeks to address the problem of data heterogeneity, and [19, 20], which seek to train personalized models at the worker nodes.

6.2.1.1 Convergence Analysis of Elastic Averaging SGD

Now let us understand how the convergence of elastic-averaging SGD depends on α, τ, m and other parameters. The original paper [17] only gives an analysis of EASGD with 1 local update and for quadratic objective functions. We present here a general analysis for non-convex objective functions, which follows from the Cooperative SGD framework proposed in [5]. Cooperative SGD gives unified convergence analysis of local-update SGD, elastic-averaging and other variants of distributed SGD.

Theorem 6.2 (Convergence of elastic-averaging SGD) *For a L-smooth function, and starting point is $\mathbf{w}_1$ then $F(\mathbf{w}_t)$ after t iterations of elastic averaging SGD is bounded as*

$$
\mathbb{E}\left[\frac{1}{t} \sum_{k=1}^{t} \|\nabla F(\mathbf{y}_k)\|^2 \right] \leq \frac{2(m+1)\left[F(\mathbf{w}_1) - F_{inf} \right]}{\eta m t} + \frac{\eta L \sigma^2}{(m+1)b}
$$

$$
+ \frac{\eta^2 L^2 \sigma^2}{b} \left(\frac{1+\zeta^2}{1-\zeta^2} \tau - 1 \right) \tag{6.62}
$$

where $\mathbf{y}_k = (\sum_{i=1}^{m} \mathbf{w}_k^{(i)} + \mathbf{z}_k)/(m+1)$ and $\zeta = \max(|1-\alpha|, |1-(m+1)\alpha|)$, which implies that

$$
\zeta = \begin{cases} (m+1)\alpha - 1, & \frac{2}{m+2} < \alpha \leq \frac{2}{m+1} \\ 1 - \alpha, & 0 \leq \alpha \leq \frac{2}{m+2} \end{cases}. \tag{6.63}
$$

To prove the above result, [5] defines a model matrix that concatenates the worker models and the anchor model, and a mixing matrix that determines how consensus is performed between these models after each round of local SGD updates. The parameter ζ is the spectral gap of the mixing matrix, and it depends on the elasticity parameter α that controls the tightness of the coupling between the anchor model and each of the worker models. From the convergence analysis presented above, we can infer the optimal value of α, the elasticity parameter. Observe that minimizing the value of the spectral gap ζ will minimize the error floor at convergence (the third term in (6.62)). When $\alpha = \frac{2}{m+2}$, one can get the minimal value of ζ, which equals to $1 - \alpha = (m+1)\alpha - 1 = \frac{m}{m+2}$. When α is set to the optimal value $2/(m+2)$, the error of EASGD can be bounded as follows:

$$\mathbb{E}\left[\frac{1}{t}\sum_{k=1}^{t}\|\nabla F(\mathbf{y}_k)\|^2\right] \leq \frac{2(m+1)\left[F(\mathbf{w}_1) - F_{\inf}\right]}{\eta m t} + \frac{\eta L \sigma^2}{b(m+1)}$$

$$+ \frac{1}{2b}\eta^2 L^2 \sigma^2 \left[\frac{(m+1)^2 + 1}{2(m+1)}\tau - 1\right] \qquad (6.64)$$

where $\mathbf{y}_k = (\sum_{i=1}^{m}\mathbf{w}_k^{(i)} + \mathbf{z}_k)/(m+1)$.

The optimal value of $\alpha = 2/(m+2)$ matches with the experimental results on training the VGG-16 network with the CIFAR-10 dataset shown in Fig. 6.9a, which shows the training loss versus iterations convergence of EASGD with different values of α.

6.2.1.2 Overlapping Communication and Computation

The anchor models $\mathbf{z}_k$ can be thought of as slightly stale versions of the average model, since they remain the same while worker nodes conduct local updates, that is, $\mathbf{z}_{j\tau} = \mathbf{z}_{j\tau-1} = \cdots = \mathbf{z}_{(j-1)\tau+1}$ for $j \geq 1$. Observe that according to the update rule (6.59), the worker nodes only need $\mathbf{z}_{(j-1)\tau+1}$ before the model-averaging step from $\mathbf{w}_{j\tau}$ to $\mathbf{w}_{j\tau+1}$. So, the anchor models can perform averaging of the previous round of local updates and broadcast their new version while the workers perform the next set of local updates. Thus, *anchor models allow the local computation to overlap with inter-node communication and enabling non-blocking execution*, as illustrated in Fig. 6.10. In Fig. 6.9b, we show experimental results demonstrating that this overlap of the the update of auxiliary variable and workers' local computation, directly reduces about 50% training time. Although the elastic averaging SGD algorithm proposed in [17] allows this natural overlap between communication and computation, the authors did not take advantage of it in a non-blocking implementation to reduce the training time.

6.2.2 Overlap Local SGD

Overlap local SGD is a variation of elastic averaging SGD, again using an anchor model $\mathbf{z}$, but updating it in a slightly different way. The update rules of overlap local SGD are:

$$\mathbf{w}_{k+1}^{(i)} = \begin{cases} \mathbf{w}_k^{(i)} - \eta g_i(\mathbf{w}_k^{(i)}; \xi_k^{(i)}) - \alpha(\mathbf{w}_k^{(i)} - \eta g_i(\mathbf{w}_k^{(i)}; \xi_k^{(i)}) - \mathbf{z}_k) & k \bmod \tau = 0 \\ \mathbf{w}_k^{(i)} - \eta g_i(\mathbf{w}_k^{(i)}; \xi_k^{(i)}) & \text{otherwise} \end{cases} \qquad (6.65)$$

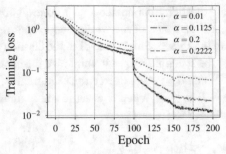 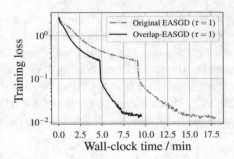

(a) Average training loss of local models.   (b) Benefit of using an anchor model: overlapping communication and computation.

Fig. 6.9 EASGD training on CIFAR-10 with VGG-16. Since there are 8 worker nodes, the best value of α is $2/(m+2) = 0.2$, which performs better than the empirical choice $\alpha = 0.9/m = 0.1125$ suggested in [17]. The best choice of α yields the lowest training loss and the least discrepancies between worker and the anchor models

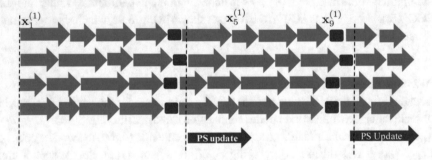

Fig. 6.10 Illustration of the runtime of elastic averaging SGD. Observe that the updates to the anchor model at the parameter server can occur in parallel to the local computation at the worker nodes

$$\mathbf{z}_{k+1} = \begin{cases} \frac{1}{m}\sum_{i=1}^{m}\mathbf{w}_{k+1}^{(i)} & k \bmod \tau = 0 \\ \mathbf{z}_k & \text{otherwise} \end{cases} \qquad (6.66)$$

where α is a tunable parameter. A larger value of α means that the locally trained model $\mathbf{w}^{(i)}$ is pulled closer to the anchor model $\mathbf{z}$. Meanwhile, another thread (or process) on each node will synchronize the current local models in parallel and store the average value into the anchor model as per (6.66). The updates in Eq. (6.65) do not involve any communication, because each node has one local copy of the anchor model. Right after pulling back, nodes will start next round of local updates immediately. It also offers the bonus benefit of straggler mitigation—the slowest node can be up to τ local updates behind the fastest node (Fig. 6.11).

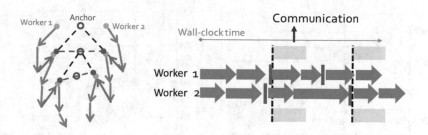

Fig. 6.11 Illustration of Overlap SGD in the parameter space and in terms of wall-clock time

6.2.2.1 Convergence Analysis of Overlap Local SGD

The convergence analysis of overlap local SGD is given by the following theorem.

Theorem 6.3 *Suppose all local models and anchor model are initialized at the same point* $\mathbf{w}_0^{(i)} = \mathbf{z}_0$ *for all* $i \in \{1, \ldots, m\}$. *Under Assumptions 1 to 4, if the learning rate is set as* $\eta = \frac{1}{L}\sqrt{\frac{m}{t}}$, *and the total iterations* t *satisfies* $t \geq 60m\tau^2/\alpha^2$, *then we have*

$$\frac{1}{t}\sum_{k=0}^{t-1} \mathbb{E}\left[\|\nabla F(\mathbf{y}_k)\|^2\right] \leq \frac{4L[F(\mathbf{y}_0) - F_{inf}]}{(1-\alpha)\sqrt{mt}} + \frac{2(1-\alpha)\sigma^2}{b\sqrt{mt}} + \frac{2m\sigma^2}{bt}\left[\frac{2}{(2-\alpha)\alpha}\tau - 1\right]$$

$$+ \frac{2m\tau^2\kappa^2}{\alpha^2 t} \qquad (6.67)$$

$$= \mathcal{O}\left(\frac{1}{\sqrt{mt}}\right) + \mathcal{O}\left(\frac{1}{t}\right). \qquad (6.68)$$

where $\mathbf{y}_k = (1-\alpha)\sum_{i=1}^{m} \mathbf{w}_k^{(i)}/m + \alpha\mathbf{z}_k$ *and* F_{inf} *is the lower bound of the objective value.*

Due to space limitation, please refer [22] for the detailed proof. The proof technique is inspired by [5]. Theorem 6.3 also shows that when the learning rate is configured properly and the total iterations t is sufficiently large, the error bound will be dominated by $1/\sqrt{mt}$, matching the same rate as fully synchronous SGD.

In Fig. 6.12 we compare the performance of Overlap Local-SGD with Elastic Averaging SGD with momentum (EAMSGD) proposed in [17] and an algorithm called CoCoD-SGD proposed in [21]. We observe the overlap local SGD is faster, both in terms of the error versus iterations and the error versus wallclock time convergence. In particular, the improvement over EAMSGD is due to the fact that in EASGD, there is an inertia in the anchor model updates, where the previous anchor model gets in weight $(1 - m\alpha)$ in (6.60). However, overlap local SGD removes this inertia in the anchor model update, as we see in (6.66).

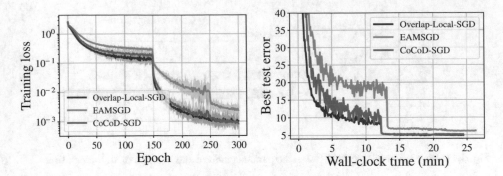

Fig. 6.12 Comparison to CoCoD-SGD proposed in [21] and elastic averaging SGD with momentum (EAMSGD) proposed in [17]. In all algorithms, the number of local updates $\tau = 2$. Overlap Local SGD slightly improves the loss-versus-iterations convergence of CoCoD-SGD

Summary

In this chapter, we consider communication-efficient distributed SGD algorithms that try to reduce the frequency of communication between the worker nodes and the parameter server. These algorithms are especially relevant in bandwidth-limited settings such as federated learning. We first discussed local-update SGD, which allows workers to perform τ local updates before the parameter server averages the model versions at the m workers. We showed that there is a trade-off between between the error convergence and runtime in terms of τ and m, and we studied an adaptive communication strategy that changes τ during training process in order to achieve a favorable error-runtime convergence. Finally, we discussed elastic averaging and overlap local SGD, which are variants of local-update SGD that modify the update rule to allow an overlap between the local computation at the workers and the communication with the parameter server.

Problems

1. For a linear regression problem, implement local-update SGD with a fixed learning rate η and plot the residual sum of squares error versus the number of communication rounds with different values of τ. How does the error floor change as a function of τ?
2. By simulating and plotting the expected runtime per iteration of local update SGD for Pareto distributed $X \sim \text{Pareto}(x_m, \alpha)$, compare its scaling with m for different values of the shape parameter τ and α.

References

1. P. Kairouz, H. B. McMahan, B. Avent, A. Bellet, M. Bennis, A. N. Bhagoji, K. Bonawitz, Z. Charles, G. Cormode, R. Cummings, R. G. L. D'Oliveira, S. E. Rouayheb, D. Evans, J. Gardner, Z. Garrett, A. Gascon, B. Ghazi, P. B. Gibbons, M. Gruteser, Z. Harchaoui, C. He, L. He, Z. Huo, B. Hutchinson, J. Hsu, M. Jaggi, T. Javidi, G. Joshi, M. Khodak, J. Konecny, A. Korolova, F. Koushanfar, S. Koyejo, T. Lepoint, Y. Liu, P. Mittal, M. Mohri, R. Nock, A. Ozgur, R. Pagh, M. Raykova, H. Qi, D. Ramage, R. Raskar, D. Song, W. Song, S. U. Stich, Z. Sun, A. T. Suresh, F. Tramer, P. Vepakomma, J. Wang, L. Xiong, Z. Xu, Q. Yang, F. X. Yu, H. Yu, and S. Zhao, "Advances and open problems in federated learning," *Foundations and Trends in Machine Learning*, Jun. 2021. [Online]. Available: https://arxiv.org/abs/1912.04977

2. H. B. McMahan, E. Moore, D. Ramage, S. Hampson, and B. A. y Arcas, "Communication-Efficient Learning of Deep Networks from Decentralized Data," *International Conference on Artificial Intelligenece and Statistics (AISTATS)*, Apr. 2017. [Online]. Available: https://arxiv.org/abs/1602.05629

3. J. Wang, Z. Charles, Z. Xu, G. Joshi, H. B. McMahan, B. A. y Arcas, M. Al-Shedivat, G. Andrew, S. Avestimehr, K. Daly, D. Data, S. Diggavi, H. Eichner, A. Gadhikar, Z. Garrett, A. M. Girgis, F. Hanzely, A. Hard, C. He, S. Horvath, Z. Huo, A. Ingerman, M. Jaggi, T. Javidi, P. Kairouz, S. Kale, S. P. Karimireddy, J. Konecny, S. Koyejo, T. Li, L. Liu, M. Mohri, H. Qi, S. J. Reddi, P. Richtarik, K. Singhal, V. Smith, M. Soltanolkotabi, W. Song, A. T. Suresh, S. U. Stich, A. Talwalkar, H. Wang, B. Woodworth, S. Wu, F. X. Yu, H. Yuan, M. Zaheer, M. Zhang, T. Zhang, C. Zheng, C. Zhu, and W. Zhu, "A field guide to federated optimization," 2021. [Online]. Available: https://arxiv.org/abs/2107.06917

4. K. Bonawitz, H. Eichner, W. Grieskamp, D. Huba, A. Ingerman, V. Ivanov, C. Kiddon, J. Konecny, S. Mazzocchi, H. B. McMahan, T. V. Overveldt, D. Petrou, D. Ramage, and J. Rose-lander, "Towards Federated Learning at Scale: System Design," *SysML*, Apr. 2019. [Online]. Available: https://www.sysml.cc/doc/2019/193.pdf

5. J. Wang and G. Joshi, "Cooperative sgd: A unified framework for the design and analysis of local-update sgd algorithms," *Journal of Machine Learning Research*, vol. 22, no. 213, pp. 1–50, 2021. [Online]. Available: http://jmlr.org/papers/v22/20-147.html

6. S. U. Stich, "Local SGD converges fast and communicates little," *arXiv preprint arXiv:1805.09767*, 2018.

7. K. Simonyan and A. Zisserman, "Very deep convolutional networks for large-scale image recognition," in *International Conference on Learning Representations (ICLR)*, 2015.

8. A. Krizhevsky and G. Hinton, "Learning multiple layers of features from tiny images," *Technical report, University of Toronto*, 2009.

9. B. Recht, C. Re, S. Wright, and F. Niu, "Hogwild: A lock-free approach to parallelizing stochastic gradient descent," in *Proceedings of the International Conference on Neural Information Processing Systems*, 2011, pp. 693–701.

10. J. Dean, G. S. Corrado, R. Monga, K. Chen, M. Devin, Q. V. Le, M. Z. Mao, M. Ranzato, A. Senior, P. Tucker, K. Yang, and A. Y. Ng, "Large scale distributed deep networks," in *Proceedings of the International Conference on Neural Information Processing Systems*, 2012, pp. 1223–1231.

11. J. Cipar, Q. Ho, J. K. Kim, S. Lee, G. R. Ganger, G. Gibson, K. Keeton, and E. Xing, "Solving the straggler problem with bounded staleness," in *Proceedings of the Workshop on Hot Topics in Operating Systems*, 2013.

12. Q. Ho, J. Cipar, H. Cui, J. K. Kim, S. Lee, P. B. Gibbons, G. A. Gibson, G. R. Ganger, and E. P. Xing, "More effective distributed ml via a stale synchronous parallel parameter server," in *Proceedings of the International Conference on Neural Information Processing Systems*, 2013, pp. 1223–1231.

13. S. Gupta, W. Zhang, and F. Wang, "Model accuracy and runtime tradeoff in distributed deep learning: A systematic study," in *IEEE International Conference on Data Mining (ICDM)*. IEEE, 2016, pp. 171–180.
14. J. Zhang, I. Mitliagkas, and C. Re, "Yellowfin and the art of momentum tuning," *CoRR*, vol. arXiv:1706.03471, Jun. 2017. [Online]. Available: http://arxiv.org/abs/1706.03471
15. S. Dutta, G. Joshi, S. Ghosh, P. Dube, and P. Nagpurkar, "Slow and Stale Gradients Can Win the Race: Error-Runtime Trade-offs in Distributed SGD," *International Conference on Artificial Intelligence and Statistics (AISTATS)*, Apr. 2018. [Online]. Available: https://arxiv.org/abs/1803.01113
16. H. Wang, S. Guo, and R. Li, "Osp: Overlapping computation and communication in parameter server for fast machine learning," in *Proceedings of the 48th International Conference on Parallel Processing*, 2019, pp. 1–10.
17. S. Zhang, A. E. Choromanska, and Y. LeCun, "Deep learning with elastic averaging SGD," in *NIPS'15 Proceedings of the 28th International Conference on Neural Information Processing Systems*, 2015, pp. 685–693.
18. X. Li, K. Huang, W. Yang, S. Wang, and Z. Zhang, "On the convergence of fedavg on non-iid data," in *International Conference on Learning Representations (ICLR)*, Jul. 2020. [Online]. Available: https://arxiv.org/abs/1907.02189
19. F. Hanzely and P. Richtárik, "Federated learning of a mixture of global and local models," *arXiv preprint* arXiv:2002.05516, 2020.
20. T. Li, S. Hu, A. Beirami, and V. Smith, "Ditto: Fair and robust federated learning through personalization," in *Proceedings of the 38th International Conference on Machine Learning*, ser. Proceedings of Machine Learning Research, vol. 139. PMLR, 18–24 Jul 2021, pp. 6357–6368. [Online]. Available: https://proceedings.mlr.press/v139/li21h.html
21. S. Shen, L. Xu, J. Liu, X. Liang, and Y. Cheng, "Faster distributed deep net training: Computation and communication decoupled stochastic gradient descent," in *IJCAI*, 2019.
22. "Overlap Local-SGD: An Algorithmic Approach to Hide Communication Delays in Distributed SGD," in *Proceedings of the International Conference on Acoustics, Speech, and Signal Processing (ICASSP)*.

Quantized and Sparsified Distributed SGD 7

The communication delay in sending and receiving gradients or model updates between the worker nodes and the parameter server can be significantly, especially in networks with limited bandwidth and high latency on the communication links. In Chap. 6 we studied a class of communication-efficient distributed SGD algorithms that save the communication cost by reducing the frequency with which workers communicate with the parameter server. In this chapter, we consider an orthogonal approach to reduce communication where we reduce the number of bits transmitted every time the workers send updates to the parameter server using quantization [1–3] or sparsification techniques [4–6]. These methods can be used in the place of or in combination with the local-update SGD methods that we discussed in Chap. 6.

When we quantize or sparsify gradients, it can increase the noise variance in the aggregated gradient used by the parameter server to update the model. We will analyze how this added noise affects the convergence of distributed SGD. We will also study how quantization or sparsification affects the runtime per iteration in order to understand which system architecture would benefit the most from the use of these techniques.

7.1 Quantized SGD

Recall that the update rule of synchronous SGD is given by:

$$\mathbf{w}_{t+1} = \mathbf{w}_t - \eta \frac{1}{m} \sum_{i=1}^{m} g(\mathbf{w}_t; \xi_i) \tag{7.1}$$

For a d-dimensional model $\mathbf{w}_t$, the corresponding gradient vectors $g(\mathbf{w}_t; \xi_i)$ are also d-dimensional vectors, with each floating point element requiring 32 (or 64) bits to represent it. For high-dimensional models (large d) this cost of communicating $32d$ bits per iteration

© The Author(s), under exclusive license to Springer Nature Switzerland AG 2023
G. Joshi, *Optimization Algorithms for Distributed Machine Learning*,
Synthesis Lectures on Learning, Networks, and Algorithms,
https://doi.org/10.1007/978-3-031-19067-4_7

can be prohibitively expensive. Quantized SGD reduces this cost by using fewer bits to represent each dimension of the gradient vector.

$$\mathbf{w}_{t+1} = \mathbf{w}_t - \eta \frac{1}{m} \sum_{i=1}^{m} Q(g(\mathbf{w}_t; \xi_i)) \tag{7.2}$$

where $Q(\cdot)$ is a quantization operator that reduces the number of bits used to represent each element of the d-dimensional gradient vector.

A naive way to quantize a vector $\mathbf{v}$ is fixed quantization where we can normalize each element of the vector such that it lies in the interval $[0, 1]$. The normalized version of the j-th element is $\frac{v_j}{\|\mathbf{v}\|_2}$. Next, to quantize $\mathbf{v}$, we divide $[0, 1]$ into s intervals of size $1/s$, and map each normalized $\frac{v_j}{\|\mathbf{v}\|_2}$ to the center of the interval that it lies in. However, a problem with this strategy is the the quantization error is higher for values that are further away from the center of the interval. This causes a bias in the quantized estimate of each element, that is, $\mathbb{E}\left[Q(v_j)\right] \neq v_j$.

7.1.1   Uniform Stochastic Quantization

Instead, using *stochastic quantization* preserves the expected value of the gradient and ensures that $eQ(\mathbf{v}) = \mathbf{v}$ and minimizes the error variance. The stochastically quantized version of vector $\mathbf{v}$ is given by,

$$Q(v_j) = \|\mathbf{v}\|_2 \cdot \text{sign}(v_j) \cdot \gamma_j(\mathbf{v}, s) \tag{7.3}$$

where γ_j are independent random variables that are defined as follows. Let $0 \leq \ell < s$ be an integer such that $\frac{|v_j|}{\|\mathbf{v}\|} \in [\ell/s, (\ell+1)/s]$, the corresponding quantization interval. Then,

$$\gamma_j(\mathbf{v}, s) = \begin{cases} (\ell+1)/s & \text{with prob. } \frac{|v_j|}{\|\mathbf{v}\|}s - \ell \\ \ell/s & \text{otherwise} \end{cases} \tag{7.4}$$

See the example shown in Fig. 7.1.

The term $\text{sign}(v_j)$ needs 1 bit for each j, d bits for a d-dimensional vector $\mathbf{v}$, $\gamma_j(\mathbf{v}, s)$ takes one of $s + 1$ values and thus needs $\log_2(s + 1)$ bits for each j, and $d \log_2(s + 1)$ bits in total, $\|\mathbf{v}\|_2$ is in floating point and needs 32 bits. Thus, the total number of bits required to represent the gradient vector is:

$$b = d(1 + \log_2(s + 1)) + 32 \tag{7.5}$$

We can choose s such that this is significantly smaller than $32d$ bits required to represent the unquantized vector $\mathbf{v}$.

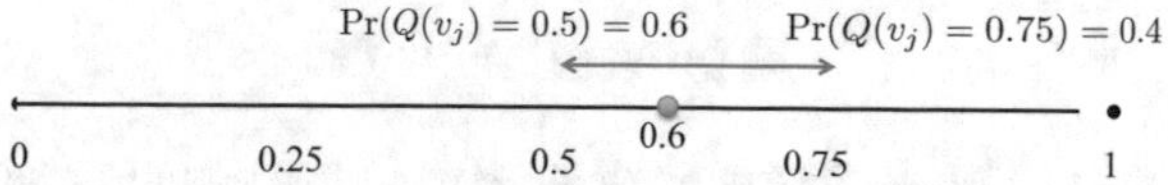

Fig. 7.1 Illustration of stochastic quantization with $s = 4$ levels

7.1.1.1 Properties of the Uniform Stochastic Quantizer

The uniform stochastic quantizer described above satisfies the following properties.

- **Unbiased Estimator**: The expected value $\mathbb{E}[Q(\mathbf{v})] = \mathbf{v}$, which follows from the definition above.
- **Bounded Estimation Error**:

$$\mathbb{E}[\|Q(\mathbf{v}) - \mathbf{v}\|_2^2] \le \min\left(\frac{d}{s^2}, \frac{\sqrt{d}}{s}\right)\|\mathbf{v}\|_2^2.$$

The proof of this claim is given in the Appendix of [1].

7.1.1.2 Extensions and Generalizations

In (7.2), we quantize the gradient computed by each worker and reduce the number of bits used to represent it. This quantization approach can be combined with local-update SGD such that each worker takes τ local updates to the model $\mathbf{w}_k$, and the model update vector $\mathbf{v} = \mathbf{w}_{\tau+k}^{(i)} - \mathbf{w}_k$ is quantized using stochastic quantization. This method is proposed in [2] and referred to as FedPAQ.

Another extension that saves the number of bits communicated is the representation of $\gamma_j(\mathbf{v}, s)$. Since it takes one of $s + 1$ values, we consider that it needs $\log_2(s + 1)$ bits to represent it. However, all the $s + 1$ values are not equally likely. Thus, we can use a variable length encoding scheme such as Huffman coding or Elias coding to encode the $\gamma_j(\mathbf{v}, s)$, and reduce the total number of transmitted bits.

7.1.2 Convergence Analysis

We want to understand how the convergence speed is affected by the number of quantization levels s.

7.1.2.1 Assumptions

- **Unbiased Gradients**: The stochastic gradient $g(\mathbf{w}; \xi)$ is an unbiased estimate of $\nabla F(\mathbf{w})$, that is,

$$\mathbb{E}_\xi[g(\mathbf{w}; \xi)] = \nabla F(\mathbf{w}) \tag{7.6}$$

- **Bounded Variance**: The stochastic gradient $g(\mathbf{w}; \xi)$ has bounded variance, that is,

$$\mathrm{Var}(g(\mathbf{w}; \xi)) \leq \frac{\sigma^2}{b} \tag{7.7}$$

$$\mathbb{E}_\xi[\|g(\mathbf{w}; \xi)\|^2] \leq \|\nabla F(\mathbf{w})\|^2 + \frac{\sigma^2}{b} \tag{7.8}$$

7.1.2.2 Error Analysis

By combining the variance upper bound with the bounded estimation property of the stochastic quantizer, we can show that

$$\mathbb{E}_Q[\|Q(g(\mathbf{w}; \xi)) - g(\mathbf{w}; \xi)\|_2^2] \leq \min\left(\frac{d}{s^2}, \frac{\sqrt{d}}{s}\right) \|g(\mathbf{w}; \xi)\|_2^2 \tag{7.9}$$

$$\mathbb{E}_Q[\|Q(g(\mathbf{w}; \xi))\|_2^2] \leq \|g(\mathbf{w}; \xi))\|_2^2 + \min\left(\frac{d}{s^2}, \frac{\sqrt{d}}{s}\right) \|g(\mathbf{w}; \xi)\|_2^2 \tag{7.10}$$

$$\mathbb{E}_\xi[\mathbb{E}_Q[\|Q(g(\mathbf{w}; \xi))\|_2^2]] \leq \mathbb{E}_\xi[\|g(\mathbf{w}; \xi))\|_2^2]$$
$$+ \min\left(\frac{d}{s^2}, \frac{\sqrt{d}}{s}\right) \mathbb{E}_\xi[\|g(\mathbf{w}; \xi)\|_2^2] \tag{7.11}$$

$$\mathbb{E}[\|Q(g(\mathbf{w}; \xi))\|_2^2] \leq \left(1 + \min\left(\frac{d}{s^2}, \frac{\sqrt{d}}{s}\right)\right) \mathbb{E}_\xi[\|g(\mathbf{w}; \xi)\|_2^2] \tag{7.12}$$

$$\leq \left(1 + \min\left(\frac{d}{s^2}, \frac{\sqrt{d}}{s}\right)\right) \left(\|\nabla F(\mathbf{w})\|^2 + \frac{\sigma^2}{b}\right) \tag{7.13}$$

Using the above bound on the second moment of the quantized gradient, we can obtain a convergence analysis of quantized SGD. The analysis generalizes the convergence analysis of SGD for non-convex objective functions given in Theorem 3.4.

Theorem 7.1 (Convergence of Quantized Distributed SGD *For a L-smooth function, with* $\eta \leq \dfrac{1}{L\left(1+\min\left(\frac{d}{s^2}, \frac{\sqrt{d}}{s}\right)\right)}$ *and starting point is* $\mathbf{w}_1$, *after t iterations of quantized distributed SGD we have*

$$\mathbb{E}\left[\frac{1}{t}\sum_{k=1}^t \|\nabla F(\mathbf{w}_k)\|^2\right] \leq \frac{2(F(\mathbf{w}_1) - F_{inf})}{\eta t} + \frac{\eta L}{m}\left(1 + \min\left(\frac{d}{s^2}, \frac{\sqrt{d}}{s}\right)\right)\frac{\sigma^2}{b} \tag{7.14}$$

where $\mathbf{w}_k$ *denotes the averaged model at the k-th iteration.*

The second term increases if we use fewer quantization levels s. Thus, the error versus iterations convergence becomes worse if we reduce s.

Proof Starting with the Lipschitz smoothness of the objective function we have

$$F(\mathbf{w}_{k+1}) - F(\mathbf{w}_k) \leq \nabla F(\mathbf{w}_k)^\top (\mathbf{w}_{k+1} - \mathbf{w}_k) + \frac{L}{2}\|\mathbf{w}_{k+1} - \mathbf{w}_k\|^2 \tag{7.15}$$

$$\leq \nabla F(\mathbf{w}_k)^\top (-\frac{\eta}{m}\sum_{i=1}^m Q(g(\mathbf{w}_k, \xi_i))) + \frac{L}{2}\| -\eta\frac{1}{m}\sum_{i=1}^m Q(g(\mathbf{w}_k, \xi_i))\|^2 \tag{7.16}$$

$$\leq -\frac{\eta}{m}\left(\sum_{i=1}^m \nabla F(\mathbf{w}_k)^\top Q(g(\mathbf{w}_k, \xi_i))\right) + \frac{\eta^2 L}{2m^2}\|\sum_{i=1}^m Q(g(\mathbf{w}_k, \xi_i))\|^2 \tag{7.17}$$

Taking expectation on both sides with respect to the stochasticity ξ_t in the t-th iteration,

$$\mathbb{E}\left[F(\mathbf{w}_{k+1})\right] - F(\mathbf{w}_k) \leq -\frac{\eta}{m}\left(\sum_{i=1}^m \nabla F(\mathbf{w}_k)^\top \mathbb{E}\left[Q(g(\mathbf{w}_k, \xi_i))\right]\right)$$
$$+ \frac{\eta^2 L}{2m^2}\sum_{i=1}^m \mathbb{E}\left[\|Q(g(\mathbf{w}_k, \xi_i))\|^2\right] \tag{7.18}$$

Using the unbiased gradient assumption $\mathbb{E}_\xi[Q(g(\mathbf{w}; \xi))] = \nabla F(\mathbf{w})$ and the bound given by (7.13) we have

$$\mathbb{E}\left[F(\mathbf{w}_{k+1})\right] - F(\mathbf{w}_k) \tag{7.19}$$

$$\leq -\eta\|\nabla F(\mathbf{w}_k)\|^2 + \frac{\eta^2 L}{2m}\left(1 + \min\left(\frac{d}{s^2}, \frac{\sqrt{d}}{s}\right)\right)\left(\|\nabla F(\mathbf{w}_k)\|^2 + \frac{\sigma^2}{b}\right) \tag{7.20}$$

$$\leq \eta\left(1 - \frac{\eta L}{2m}\left(1 + \min\left(\frac{d}{s^2}, \frac{\sqrt{d}}{s}\right)\right)\right)(-\|\nabla F(\mathbf{w}_k)\|_2^2) +$$
$$\frac{\eta^2 L}{2m}\left(1 + \min\left(\frac{d}{s^2}, \frac{\sqrt{d}}{s}\right)\right)\frac{\sigma^2}{b} \tag{7.21}$$

$$\leq \frac{\eta}{2}(-\|\nabla F(\mathbf{w}_k)\|_2^2) + \frac{\eta^2 L}{2m}\left(1 + \min\left(\frac{d}{s^2}, \frac{\sqrt{d}}{s}\right)\right)\frac{\sigma^2}{b} \tag{7.22}$$

where we get the last inequality by using the assumption that $\eta \leq \frac{1}{L\left(1+\min\left(\frac{d}{s^2}, \frac{\sqrt{d}}{s}\right)\right)}$. Summing the two sides from $k = 1, \ldots, t$ and dividing by t we get

$$\frac{F_{inf} - F(\mathbf{w}_1)}{t} \le \frac{\eta}{2t}(-\sum_{k=1}^{t} \|\nabla F(\mathbf{w}_k)\|_2^2) + \frac{\eta^2 L}{2m} \left(1 + \min\left(\frac{d}{s^2}, \frac{\sqrt{d}}{s}\right)\right) \frac{\sigma^2}{b} \quad (7.23)$$

$$\frac{1}{t} \sum_{k=1}^{t} \|\nabla F(\mathbf{w}_k)\|_2^2 \le \frac{2(F(\mathbf{w}_1) - F_{inf})}{\eta t} + \frac{\eta L}{m} \left(1 + \min\left(\frac{d}{s^2}, \frac{\sqrt{d}}{s}\right)\right) \frac{\sigma^2}{b} \quad (7.24)$$

In FedPAQ, quantization is combined with local updates in order to further improve the communication efficiency. The convergence of quantized local-update SGD can be analyzed as follows, where we quantity the effect of the number of local updates τ.

Theorem 7.2 (Convergence of Quantized Local-update SGD) *For a L-smooth function,* $\eta L \left(1 + \min\left(\frac{d}{s^2}, \frac{\sqrt{d}}{s}\right)\right) + \eta^2 L^2 \tau(\tau - 1) \le 1$, *and if the starting point is* $\mathbf{w}_1$ *then* $F(\mathbf{w}_t)$ *after t iterations of local update SGD is bounded as*

$$\mathbb{E}\left[\frac{1}{t}\sum_{k=1}^{t} \|\nabla F(\mathbf{w}_k)\|^2\right] \le \frac{2\left[F(\mathbf{w}_1) - F_{inf}\right]}{\eta t} + \frac{\eta L \sigma^2}{mb}\left(1 + \min\left(\frac{d}{s^2}, \frac{\sqrt{d}}{s}\right)\right)$$
$$+ \frac{\eta^2 L^2 \sigma^2(\tau - 1)}{b} \quad (7.25)$$

where $\mathbf{w}_k = \sum_{i=1}^{m} \mathbf{w}_k^{(i)}/m$ *denotes the averaged model at the k-th iteration. This is a virtual sequence of iterates that is equal to the global model after every τ iterations when $(k - 1)$ mod $\tau = 0$.*

The proof follows similarly as the proof of local-update SGD, expect with σ^2/b replaced by $\left(1 + \min\left(\frac{d}{s^2}, \frac{\sqrt{d}}{s}\right)\right) \frac{\sigma^2}{b}$ at the time when the sparsified updates are averaged by the parameter server.

7.1.3　Runtime Analysis

The runtime per iteration of quantized SGD is proportional to the number of bits communicated per iteration (or per round, in the case of quantized local SGD). Let us first determine the number of bits communicated, and then analyze how it affects the communication delay per iteration.

7.1.3.1 Bits Communicated Per Iteration

For synchronous SGD, the number of bits communication per iteration by each worker is:

$$B_{sync} = 32d \quad (7.26)$$

With quantized SGD with s quantization levels, the number of bits communicated per iteration is:

$$B_{Quant} = d(1 + \log_2(s + 1)) + 32 \tag{7.27}$$

Instead of just computing and sending gradients, if the workers perform τ local updates each, the number of bits communicated get amortized across τ iterations. Thus, $B_{local} = \frac{32d}{\tau}$ for local-update SGD. For FedPAQ with τ local updates and s quantization levels used for each update, the number of bits communicated per iteration is $B_{FedPAQ} = \frac{d(1+\log_2(s+1))+32}{\tau}$.

7.1.3.2 Communication Time per Iteration

If we had to choose between one of the two ways of reducing communication, (1) local-update SGD, which reduces the frequency of communication, and (2) quantized SGD, which reduces the number of bits communicated, what strategy would be preferable? The answer depends on how the link latency and bandwidth. The communication delay can be specified in terms of two parameters: (1) the link latency D, which is independent of the size of the transmitted message, and (2) the per-bit communication delay Δ, which is inversely proportional to the bandwidth of the link. The communication time per iteration of local-update SGD is given by:

$$C_{local} = \frac{D + (32d)\Delta}{\tau} \tag{7.28}$$

The communication time per iteration of quantized SGD is given by:

$$C_{quant} = D + (d(1 + \log_2(s + 1)) + 32)\Delta. \tag{7.29}$$

7.1.3.3 Runtime per Iteration

If we consider that the time to perform one gradient computation or one local update is a constant Y, then the runtimes per iteration of quantized SGD and local-update SGD are given by

$$T_{local} = Y + \frac{D + (32d)\Delta}{\tau} \tag{7.30}$$

$$T_{quant} = Y + D + (d(1 + \log_2(s + 1)) + 32)\Delta. \tag{7.31}$$

From the above equations, it is evident that in bandwidth-limited scenarios where the per-bit delay Δ is larger than D or for high-dimensional models where d is large, quantized SGD is better than local-update SGD. However, if the communication delay is dominated by the link latency D, then local-update SGD is better. FedPAQ, which proposes quantized local SGD, combines the advantages of these methods. However the best choice of τ and the number of quantization levels s is governed by the convergence analysis in Sect. 7.1.2 and the runtime analysis given above.

Fig. 7.2 Illustration of the adaptive quantization strategy

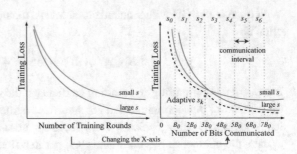

7.1.4   Adaptive Quantization

From the convergence and runtime analysis of quantized SGD and quantized local-update SGD above, it is apparent there is there a trade-off between error floor and the number of bits communicated as we vary the number of quantization levels s. For a fixed learning rate, it is better to use coarser quantization (small s) at the beginning of training, and then gradually use finer quantization (larger s) as the model comes closer to convergence, as illustrated in Fig. 7.2.

We refer to this strategy as the AdaQuant strategy [3], and obtain a heuristic schedule using the error versus communicated bits trade-off, similar to our derivation of the AdaComm strategy in Chap. 6.

To derive the schedule, we begin by considering the convergence analysis of quantized local-update SGD given in Theorem 7.2. We first assume that $\min\left(\frac{d}{s^2}, \frac{\sqrt{d}}{s}\right) = \frac{d}{s^2}$ since $\frac{\sqrt{d}}{s}$ will be smaller than $\frac{d}{s^2}$ in the extreme high-dimensional regime. Next, we replace the number of iterations t by the number of iterations that will be completed after communicating B bits in total. Since each round (consisting of τ iterations) incurs a communication cost of $(d(1 + \log_2(s + 1)) + 32)$ bits, we replace t by $B\tau/(d(1 + \log_2(s + 1)) + 32)$. By taking the derivative of the error bound with respect to s and setting it to zero, we get the following heuristic schedule:

$$s^* = \sqrt{\frac{\eta^2 L\sigma^2 \tau B \log_e(2)}{m(F(\mathbf{w}) - F^*)}} \tag{7.32}$$

In practice, the parameters such as L, σ and F^* are not unknown. Hence, to get a practically usable schedule, we first divide the number of communication bits into interval of B_0 bits each. Consider that $\mathbf{w}_l$ is the model at the beginning of the l-the interval. Assume that $F^* = 0$, compute s_0^* and s_l^* at two time intervals and take their ratio to get:

$$s_l^* = \sqrt{\frac{F(\mathbf{w}_0)}{F(\mathbf{w}_l)}} s_0^* \tag{7.33}$$

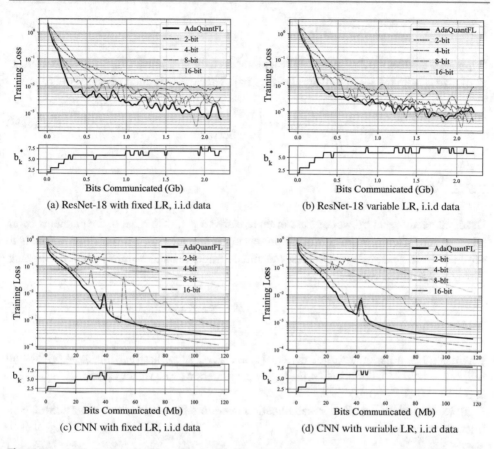

(a) ResNet-18 with fixed LR, i.i.d data

(b) ResNet-18 variable LR, i.i.d data

(c) CNN with fixed LR, i.i.d data

(d) CNN with variable LR, i.i.d data

Fig. 7.3 a, b Show that AdaQuantFL on ResNet-18 requires a fewer bits to reach a lower loss threshold, and in (**a**) AdaQuantFL reaches a loss of 0.02 in 0.3 Gb while the 2-bit method takes 1.8 Gb. For the Vanilla CNN in (**c, d**), AdaQuantFL is able to achieve the lowest error floor of 0.02 for the non-i.i.d data distribution, while other methods converge at a higher error floor. Here $b_k^* = \lceil \log_2(s_k^* + 1) \rceil$

The initial value s_0^* can be found via grid search, or set to a default $s_0^* = 2$ which we found to be a good choice in the experiments shown in Fig. 7.3.

7.2   Sparsified SGD

In quantized SGD, we quantize each element of the d-dimensional gradient or model updates to fewer bits than its floating point representation. However, it still requires communication of $O(d)$ bits, which can still be prohibitively expensive for large d. An alternative approach is to sparsify the vector and communicate only a small subset of the d dimensions of the vector [5].

7.2.1   Rand-k Sparsification

Suppose $\mathbf{v}$ denotes the d-dimensional vector that is to be sparsified. Choose a random subset Ω of size k from the set of indices $\{1, 2, \ldots, d\}$. Then the sparsified vector $S(\mathbf{v})$ is

$$S(\mathbf{v}) = \begin{cases} v_i \text{ if } i \in \Omega \\ 0 \text{ otherwise} \end{cases} \tag{7.34}$$

7.2.2   Top-k Sparsification

Gradient vectors sent by workers are often already sparse, with most of the elements being close to zero. Hence, a better way of preserving information in the sparsified vector is to send the k elements with the highest magnitude rather than k randomly chosen elements. Then the sparsified vector $S(\mathbf{v})$ is

$$S(\mathbf{v}) = \begin{cases} v_{\pi(i)} \text{ if } 1 \leq i \leq k \\ 0 \text{ otherwise} \end{cases} \tag{7.35}$$

where $\pi(i)$ is a permutation of the set $\{1, 2, \ldots, d\}$ such that $|v_{\pi(i)}| \geq |v_{\pi(i+1)}|$ for all $i = 1, 2, \ldots, d - 1$. For instance, the top-2 sparsification of the vector $\mathbf{v}[1, 5, -1, -3, -2]^\top$ is $S(\mathbf{v}) = [0, 5, 0, -3, 0]^\top$.

Both the Rand-k and Top-k sparsification operators described above satisfy the k contraction property:

$$\mathbb{E}\left[\|\mathbf{v} - S(\mathbf{v})\|^2\right] \leq \left(1 - \frac{k}{d}\right) \|\mathbf{v}\|^2 \text{ for all } \mathbf{v} \in \mathbb{R}^d \tag{7.36}$$

The proof is given in the Appendix of [5], and we restate it here. For the rand-k estimator, the above inequality, in fact, holds with an equality because

$$\mathbb{E}\left[\|\mathbf{v} - S(\mathbf{v})\|^2\right] = \sum_{j=1}^{d} \mathbb{E}\left[(\mathbf{v} - S(\mathbf{v}))_j^2\right] \tag{7.37}$$

$$= \sum_{j=1}^{d} \left(v_j^2 \times \left(1 - \frac{k}{d}\right) + 0 \times \left(\frac{k}{d}\right)\right) \tag{7.38}$$

$$= \left(1 - \frac{k}{d}\right) \|\mathbf{v}\|^2 \tag{7.39}$$

For the top-k estimator, by its definition, the value of $\mathbb{E}\left[\|\mathbf{v} - S(\mathbf{v})\|^2\right]$ is less than the corresponding value for the rand-k estimator.

7.2.3   Rand-k Sparsified Distributed SGD

Next, let us see how rand-k sparsification can be used in distributed SGD with m worker nodes. Each iteration proceeds as follows:

1. The parameter server sends the current version of the model $\mathbf{w}_k$ to the m worker nodes
2. Each worker computes the gradient $g(\mathbf{w}_k; \xi_i)$ using one mini-batch of data sampled from its local dataset $\mathcal{D}_i$.
3. Each worker then sparsifies the gradient using rand-k sparsification. Only these k elements communicate the k elements and their corresponding indices.
4. The parameter server aggregates the sparsified gradients and updates the model weights as:

$$\mathbf{w}_{k+1} = \mathbf{w}_k - \frac{\eta}{m} \sum_{i=1}^{m} \frac{d}{k} S(g(\mathbf{w}_k; \xi_i)) \tag{7.40}$$

The scaling factor d/k ensures that the aggregated gradient is an unbiased estimate of the full gradient:

$$\mathbb{E}\left[\frac{1}{m} \sum_{i=1}^{m} \frac{d}{k} S(g(\mathbf{w}_k; \xi_i)) \right] = \nabla F(\mathbf{w}_k) \tag{7.41}$$

where the expectation is taken over the random choice of the k indices in the sparsification step.

7.2.3.1 Convergence Analysis

We want to understand how the convergence speed is affected by the number of quantization levels s.

Assumptions

- **Unbiased Gradients**: The stochastic gradient $g(\mathbf{w}; \xi)$ is an unbiased estimate of $\nabla F(\mathbf{w})$, that is,

$$\mathbb{E}_\xi[g(\mathbf{w}; \xi)] = \nabla F(\mathbf{w}) \tag{7.42}$$

- **Bounded Variance**: The stochastic gradient $g(\mathbf{w}; \xi)$ has bounded variance, that is,

$$\text{Var}(g(\mathbf{w}; \xi)) \leq \frac{\sigma^2}{b} \tag{7.43}$$

$$\mathbb{E}_{\xi}[\|g(\mathbf{w}; \xi)\|^2] \leq \|\nabla F(\mathbf{w})\|^2 + \frac{\sigma^2}{b} \tag{7.44}$$

By combining the variance upper bound with the k-contraction property of the rand-k sparsification scheme, we can show that

$$\mathbb{E}\left[\|\frac{d}{k}S(g(\mathbf{w}_k; \xi)) - \nabla F(\mathbf{w}_k)\|^2\right] = \frac{d^2}{k^2}\mathbb{E}\left[\|S(g(\mathbf{w}_k; \xi))\|^2\right] - \|\nabla F(\mathbf{w}_k)\|^2 \tag{7.45}$$

$$\leq \frac{d^2}{k^2}\mathbb{E}\left[\|S(g(\mathbf{w}_k; \xi))\|^2\right] \tag{7.46}$$

$$\leq \frac{d}{k}\mathbb{E}\left[\|g(\mathbf{w}_k; \xi\|^2\right] \tag{7.47}$$

$$\leq \frac{d}{k}\left(\|\nabla F(\mathbf{w}_k)\|^2 + \frac{\sigma^2}{b}\right) \tag{7.48}$$

Using this variance bound, we can analyze the convergence of rand-k sparsified distributed SGD as follows.

Theorem 7.3 (Convergence of Rand-k- Sparsified Distributed SGD) *For a L-smooth function, with $\eta \leq \frac{k}{Ld}$ small enough and starting point is $\mathbf{w}_1$, after t iterations of rand-k sparsified distributed SGD we have*

$$\mathbb{E}\left[\frac{1}{t}\sum_{k=1}^{t}\|\nabla F(\mathbf{w}_k)\|^2\right] \leq \frac{2(F(\mathbf{w}_1) - F_{inf})}{\eta t} + \frac{\eta L}{m}\frac{d}{k}\frac{\sigma^2}{b} \tag{7.49}$$

where $\mathbf{w}_k$ denotes the averaged model at the k-th iteration.

Observe that if we sparsify aggressively, that is, if k is much smaller than d, then it can lead to a blow-up in the error floor term.

7.2.4   Error Feedback in Sparsified SGD

The variance blow-up observed above can be mitigated by using an approach called error feedback, which accumulates the sparsification error in memory at each worker and uses it to reduce the error between the aggregated sparsified gradient, which serves as an approximation of the full gradient $\nabla F(\mathbf{w})$.

Each worker maintains a memory buffer $\mathbf{y}_k^{(i)}$, a vector of dimension d, which is initialized to the all-zero vector at the beginning of training. In every iteration of sparsified SGD, the worker computes a stochastic gradient $g(\mathbf{w}_k; \xi_i)$ using a mini-batch of b samples drawn from its local dataset. Instead of simply sparsifying this gradient, it will compute the sparisifed gradient and update the memory buffer as follows:

$$\tilde{v}^{(i)}(\mathbf{w}_k) = \frac{d}{k} S(\mathbf{y}_k^{(i)} + \eta g(\mathbf{w}_k; \xi_i)) \tag{7.50}$$

$$\mathbf{y}_{k+1}^{(i)} = \mathbf{y}_k^{(i)} + \eta g(\mathbf{w}_k; \xi_i) - \tilde{v}^{(i)}(\mathbf{w}_k) \tag{7.51}$$

Now the parameter server receives the vectors $\tilde{v}^{(i)}(\mathbf{w}_k)$ for all $i = 1, \ldots, m$ and updates the model $\mathbf{w}_k$ as follows:

$$\mathbf{w}_{k+1} = \mathbf{w}_k - \frac{1}{m} \sum_{i=1}^{m} \tilde{v}^{(i)}(\mathbf{w}_k) \tag{7.52}$$

We can show that the aggregated gradient $\frac{1}{m} \sum_{i=1}^{m} \tilde{v}^{(i)}(\mathbf{w}_k)$ is an unbiased estimate of the stochastic gradient $\frac{1}{m} \sum_{i=1}^{m} \eta g(\mathbf{w}_k; \xi_i)$, (see Problem 1 below) which in turn is an unbiased estimate of the true gradient $\eta \nabla F(\mathbf{w}_k)$.

However, it has lower variance than the variance of the aggregated gradient in the case of Rand-k sparsified without memory. Here is an intuitive explanation why this is true. When we sparsify the gradient vector $g(\mathbf{w}_k; \xi_i)$ as $\frac{d}{k} S(g(\mathbf{w}_k; \xi_i))$, the $d - k$ indices in the sparsified vector are equal to zero and the information at these indices is lost. The memory buffer $\mathbf{y}_k^{(i)}$ preserves this information by accumulating the error vector $g(\mathbf{w}_k; \xi_i) - \frac{d}{k} S(g(\mathbf{w}_k; \xi_i))$. In the next iteration, since the sparsification step operates on the sum of the memory buffer and the gradient, rather than the gradient by itself, the error caused by sparsification can get canceled out in the next iteration.

Summary

In this chapter, we studied two compression techniques, quantization and sparsification, that can help reduce the number of bits transmitted over the network every time a worker communicates with the parameter server. Quantization reduces the number of bits required to represent each element of the d-dimensional gradient vector. In contrast, sparsification retains only a subset of the d elements and sets the rest to zero. In both these techniques there is a trade-off between the error and the number of bits communicated—more aggressive quantization or sparsification cuts the communication, albeit at the cost of an increase in the overall gradient variance, which causes a higher error floor.

In order to improve the error-runtime trade-off, we proposed two techniques: (1) adaptive compression, where the compression ratio is adapted during the course of training, and (2) error feedback, where the variance in the compressed gradient is reduced by accumulating errors locally at each worker and adding them to the gradient at the time of compression. While adaptive compression of updates was discussed in the context of quantization, and error feedback was discussed in the context of sparsification, these are general strategies that can be used to improve any compression operator.

Problems

1. Prove the convergence bound given in Theorem 7.3.
2. For sparsified SGD with error feedback, prove that the aggregated gradient $\frac{1}{m} \sum_{i=1}^{m} \tilde{v}^{(i)}(\mathbf{w}_k)$ an unbiased estimate of the true gradient $\eta \nabla F(\mathbf{w}_k)$.

References

1. D. Alistarh, D. Grubic, J. Li, R. Tomioka, and M. Vojnovic, "Qsgd: Communication-efficient sgd via gradient quantization and encoding," in *Advances in Neural Information Processing Systems*, 2017, pp. 1709–1720.
2. A. Reisizadeh, A. Mokhtari, H. Hassani, A. Jadbabaie, and R. Pedarsani, "Fedpaq: A communication-efficient federated learning method with periodic averaging and quantization," in *Proceedings of the Twenty Third International Conference on Artificial Intelligence and Statistics (AISTATS)*, ser. Proceedings of Machine Learning Research, vol. 108. Online: PMLR, 26–28 Aug 2020, pp. 2021–2031. [Online]. Available: http://proceedings.mlr.press/v108/reisizadeh20a.html
3. D. Jhunjhunwala, A. Gadhikar, G. Joshi, and Y. C. Eldar, "Adaptive Quantization of Model Updates for Communication-Efficient Federated Learning," in *Proceedings of the International Conference on Acoustics, Speech, and Signal Processing (ICASSP)*, May 2021. [Online]. Available: https://arxiv.org/abs/1808.07576
4. H. Wang, S. Sievert, S. Liu, Z. Charles, D. Papailiopoulos, and S. Wright, "Atomo: Communication-efficient learning via atomic sparsification," in *Advances in Neural Information Processing Systems*, 2018, pp. 9850–9861.
5. S. U. Stich, J.-B. Cordonnier, and M. Jaggi, "Sparsified sgd with memory," in *Advances in Neural Information Processing Systems*, 2018, pp. 4447–4458.
6. D. Jhunjhunwala, A. Mallick, A. Gadhikar, S. Kadhe, and G. Joshi, "Leveraging Spatial and Temporal Correlations in Sparsified Mean Estimation," in *Proceedings on Neural Information Processing Systems (NeurIPS)*, Dec. 2021. [Online]. Available: https://arxiv.org/abs/2110.07751

Decentralized SGD and Its Variants

8

In all the distributed SGD implementations that we studied so far, namely, synchronous SGD (Chap. 4), asynchronous SGD (Chap. 5), and local-update SGD (Chap. 6) and quantized and sparsified SGD (Chap. 7), we considered a central parameter server that aggregates updates and gradients from a system of m worker nodes. However, this framework may not scale well as the number of workers m grows because of the communication with a single central server can be bottleneck. In this chapter we consider decentralized SGD training architecture where we do not have a parameter server. Instead, the worker nodes are connected by an arbitrary network topology, as illustrated in Fig. 8.1. Using a decentralized topology such as this can be viewed as a strategy to achieve spatial communication-reduction, complementary to the temporal communication reduction achieved by local-update and quantized SGD.

In the rest of this chapter, we will analyze the convergence and runtime of decentralized SGD and its variants. Then we will consider some strategies to improve this error-runtime trade-off.

8.1   Network Topology and Graph Notation

We consider a network of m worker nodes, as illustrated in Fig. 8.1. The nodes are the vertices of the network graph, and correspond to the vertex set $\mathcal{V} = \{1, 2, \ldots, m\}$.

The communication links connecting the worker nodes (vertices) are represented by an arbitrary undirected graph $\mathcal{G}$ with the edge set $\mathcal{E} \in \mathcal{V} \times \mathcal{V}$. For example, in Fig. 8.1, $\mathcal{E} = \{(1, 2), (1, 3), (1, 4), (2, 6), (3, 4), (3, 5), (3, 6), (4, 5)\}$. Each node i can only communicate with its neighbors, that is, it can communicate with node j if and only if $(i, j) \in \mathcal{E}$. We use $\mathcal{N}_i = \{j | (j, i) \in \mathcal{E}, j \in \mathcal{V}, j \neq i\}$ to denote the neighbor index set of node i. For example, in Fig. 8.1, $\mathcal{N}(3) = \{1, 4, 5, 6\}$. The degree of node i is defined as $d_i = |\mathcal{N}_i|$, and the maximal node degree is denoted by $\Delta_{\mathcal{G}} = \max_{i \in \mathcal{V}} d_i$. We assume that the graph $\mathcal{G}$ is connected, that is, each node has at least one neighbor.

© The Author(s), under exclusive license to Springer Nature Switzerland AG 2023
G. Joshi, *Optimization Algorithms for Distributed Machine Learning*,
Synthesis Lectures on Learning, Networks, and Algorithms,
https://doi.org/10.1007/978-3-031-19067-4_8

Fig. 8.1 Decentralized
topology consisting of 6 nodes

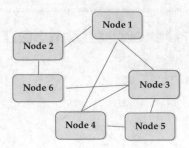

Based on the graph $\mathcal{G}$, let us now define some matrices that can be used to study the properties of the graph, in particular, the convergence of decentralized SGD algorithms run on a network of nodes connected by the graph $\mathcal{G}$.

8.1.1 Adjacency Matrix

The first matrix that the readers should know is the adjacency matrix $\mathbf{A}$, an $m \times m$ sized matrix that represents all the edges in the graph. Element $A_{i,j} = 1$ if there an edge between i and j and 0 otherwise. The diagonal elements $A_{i,i} = 0$, since there is edge from a node to itself. For example, the adjacency matrix of the graph in Fig. 8.1 is

$$\mathbf{A} = \begin{bmatrix} 0 & 1 & 1 & 1 & 0 & 0 \\ 1 & 0 & 0 & 0 & 0 & 1 \\ 1 & 0 & 0 & 1 & 1 & 1 \\ 1 & 0 & 1 & 0 & 1 & 0 \\ 0 & 0 & 1 & 1 & 0 & 0 \\ 0 & 1 & 1 & 0 & 0 & 0 \end{bmatrix} \tag{8.1}$$

By definition, the adjacency matrix is a symmetric matrix. The sum of the i-th row (or column) is equal to $d_i = |\mathcal{N}_i|$, the degree of node i, for all $i = 1, \ldots, m$.

8.1.2 Laplacian Matrix

Another way of representing the graph is called the graph Laplacian matrix $\mathbf{L}$. If $\mathbf{D}$ is a diagonal matrix consisting the node degrees, then $\mathbf{L} = \mathbf{D} - \mathbf{A}$. Its element $L_{i,j}$ are:

$$L_{i,j} = \begin{cases} \deg(v_i) & \text{if } i = j \\ -1 & \text{if } i \neq j \text{ and } (i, j) \in \mathcal{E} \\ 0 & \text{otherwise} \end{cases} \tag{8.2}$$

The Laplacian matrix of the graph in Fig. 8.1 is given by

$$
\mathbf{L} = \begin{bmatrix}
3 & -1 & -1 & -1 & 0 & 0 \\
-1 & 2 & 0 & 0 & 0 & -1 \\
-1 & 0 & 4 & -1 & -1 & -1 \\
-1 & 0 & -1 & 3 & -1 & 0 \\
0 & 0 & -1 & -1 & 2 & 0 \\
0 & -1 & -1 & 0 & 0 & 2
\end{bmatrix}
\tag{8.3}
$$

The Laplacian matrix is also symmetric, like the adjacency matrix. However, the sum of the i-th row (or column) is equal to 0 for all $i = 1, \ldots, m$. The second smallest eigenvalue $\lambda_2(\mathbf{L})$ of the Laplacian matrix is referred to as the algebraic connectivity of the graph. Larger $\lambda_2(\mathbf{L})$ implies that the graph is better connected.

8.1.3   Mixing Matrix

The adjacency and Laplacian matrices only capture information about the existence of an edge between every pair of nodes, but do not assign a weight to each edge. We now define the mixing matrix $\mathbf{M}$, which assigns weights to the edges incident on each node. These weights represent the proportion in which the information coming from neighboring nodes is mixed with local information at the node. Thus, when every node performs some local computation and shares results with its neighbors, the mixing weights can be used by each neighbor to incorporate these results into their next round of local computation. For decentralized SGD convergence discussed in this chapter we assume that the elements of the mixing matrix of an undirected graph $\mathcal{G}$ satisfy the following properties:

- The elements $M_{i,j} \geq 0$ for all $i, j \in \{1, \ldots, m\}$
- If $(i, j) \notin \mathcal{E}$, then the corresponding element $M_{i,j} = 0$
- It is symmetric, that, $\mathbf{M}^\top = \mathbf{M}$
- It is doubly stochastic, that is, the sum of every row or column is equal to 1.

8.2   Decentralized SGD

For the network topology described above, we now describe the decentralized SGD algorithm. It has been extensively studied in the field of distributed and consensus optimization [1, 2]. Recently, there has been a renewal of interest in distributed optimization owing to its applications in distributed machine learning training in a decentralized setting [3–8], without a single parameter server.

In decentralized optimization, the network of m nodes connected by the graph $\mathcal{G} = \{\mathcal{V}, \mathcal{E}\}$, seek to collaboratively train a model $\mathbf{w}$ that minimizes the following objective function:

$$F(\mathbf{w}) = \sum_{i=1}^{m} F_i(\mathbf{w}) = \sum_{i=1}^{m} \mathbb{E}_{\xi \in \mathcal{D}_i}[\ell(\mathbf{w}; \xi)] \tag{8.4}$$

where $\mathcal{D}_i$ is the local dataset at node i, and $\ell(\mathbf{w}; \xi)$ is the sample-loss for $\xi \in \mathcal{D}_i$ computed at the model parameter vector $\mathbf{w}$. The local objective function at node i is $F_i(\mathbf{w})$ and the global objective is the sum of the m local objectives.

There are two key constraints that make it challenging to directly solve the optimization problem posed in (8.4). Firstly, each node only has access to its local dataset $\mathcal{D}_i$ and the corresponding objective function $F_i(\mathbf{w})$. Secondly, each node i can only communicate with its neighboring nodes $\mathcal{N}(i)$ due to which it takes time to achieve consensus in the graph such that all nodes have the same version of the model $\mathbf{w}$. Decentralized SGD tackles the first constraint by having each node perform one local SGD step using a mini-batch of data drawn from its local dataset $\mathcal{D}_i$. The algorithm tackles the second constraint by mixing the information received from the neighbors of a node with its local information according to the mixing matrix $\mathbf{M}$.

8.2.1   The Algorithm

We now describe the classic decentralized SGD algorithm. The models $\mathbf{w}_1^{(i)}$ at the m nodes are all initialized to a common value $\mathbf{w}_1$. At the beginning of the $t + 1$-th iteration, each node i for $i = 1, 2, \ldots, m$ has its version of the model $\mathbf{w}_t^{(i)}$. The iteration proceeds as follows:

1. **Local Computation**. Each node performs one SGD update:

$$\mathbf{w}_{t+\frac{1}{2}}^{(i)} = \mathbf{w}_t^{(i)} - \eta g(\mathbf{w}_t^{(i)})$$

   Due to difference in the local datasets $\mathcal{D}_i$ across nodes, $\mathbf{w}_{t+\frac{1}{2}}^{(i)}$ will be different from each other.

2. **Inter-node Communication**. Each node then sends the updated model $\mathbf{w}_{t+\frac{1}{2}}^{(i)}$ to its neighbors $\mathcal{N}(i)$ and similarly it receives the updated models $\mathbf{w}_{t+\frac{1}{2}}^{(j)}$ for $j \in \mathcal{N}_i$ from its neighbors.

3. **Consensus**. After the communication step, each node i updates its model as per

$$\mathbf{w}_{t+1}^{(i)} = \sum_{j \in \{i, \mathcal{N}(i)\}} M_{i,j} \mathbf{w}_{t+\frac{1}{2}}^{(i)}$$

   where $\mathcal{N}(i)$ is the set of neighbors of i and $M_{i,j}$ is an element of the mixing matrix.

To write the above algorithm compactly, define matrices $\mathbf{W}_k, \mathbf{G}_k \in \mathbb{R}^{d \times m}$ that are concatenate the models and the gradients at the m workers:

$$\mathbf{W}_k = [\mathbf{w}_k^{(1)}, \dots, \mathbf{w}_k^{(m)}] \tag{8.5}$$

$$\mathbf{G}_k = [g(\mathbf{w}_k^{(1)}), \dots, g(\mathbf{w}_k^{(m)})] \tag{8.6}$$

The mixing matrix $\mathbf{M}$ determines how the local models at the workers are averaged after every iteration. Then the update rule of decentralized SGD can be written as:

$$\mathbf{W}_{k+1} = (\mathbf{W}_k - \eta \mathbf{G}_k)\mathbf{M}. \tag{8.7}$$

Synchronous distributed SGD is a special case of decentralized SGD, corresponding to a fully-connected topology and a mixing matrix $\mathbf{M}$ whose elements are all equal to $1/m$. The models at the worker nodes will be in perfect consensus after every iteration, that is, $\mathbf{w}_t^{(1)} = \mathbf{w}_t^{(2)} = \cdots = \mathbf{w}_t^{(m)}$ for all t.

8.2.2   Variants of Decentralized SGD

Next we will discuss several variants of decentralized SGD. These are only few examples drawn from the vast literature on distributed optimization.

8.2.2.1 Decentralized Local-Update SGD

By performing consensus using a sparse network topology, decentralized achieves spatial communication reduction. This strategy can be combined with temporal communication reduction, that is, reducing the communication frequency. All we need to change is that local computation step on the algorithm above, and have each of the m nodes perform $\tau > 1$ local SGD steps. The communication and consensus steps remain the same. This algorithm was proposed and analyzed in [9].

8.2.2.2 Decentralized SGD with a Time-Varying Active Topology

The basic decentralized SGD algorithm described above uses the entire network topology and the same mixing matrix in every iteration. Decentralized local-update SGD reduces the frequency of communication on each link so that the network topology is activated once after every round of τ local iterations. However, it uses the same frequency of communication on every link. Instead, [10] tunes the communication frequency according to the importance of each link in terms of maintaining the connectivity of the graph. This is achieved by decomposing the graph into disjoint matchings. Then the probability of activation of each matching is optimized such that the overall communication frequency is less than a budget, and the algebraic connectivity $\lambda_2(L)$ of the expected graph Laplacian matrix is maximized.

Connectivity-critical links are activated more frequency, whereas links between nodes that are already well-connected are activated less frequently.

8.2.2.3 Decentralized SGD on Directed Graphs

Considering a doubly stochastic and symmetric mixing matrix implies that for every pair of nodes i and j: (1) if i sends its updates to j, then it also receives an update from j, and (2) the weight assigned by node i to node j's update is equal to the weight assigned by node j to node i. These assumptions can lead to deadlocks in the implementation due to straggling nodes. To reduce the coupling of inter-node messages, we can use a directed graph, which is represented by a column stochastic matrix, to perform consensus between nodes. Based on this idea, an algorithm called Stochastic Gradient Push-sum was recently proposed in [8].

8.2.2.4 Network Gradient Tracking and Variance Reduction

Instead of having each node do one or more vanilla SGD updates, [11] proposes using variance-reduction techniques to improve the convergence of decentralized SGD. The paper proposes two algorithms Network-SVRG (based on stochastic variance reduction) and Network-DANE (based on gradient tracking) that incorporate information about previously computed gradients in each local SGD update.

8.2.2.5 Decentralized Elastic Averaging SGD

To increase consensus between a loosely connected set of workers, we can employ the idea of elastic averaging SGD, which we introduced in Chap. 6. We can do so by adding copies of an anchor model $\mathbf{z}$, a vector of the same size as the model $\mathbf{w}$, to each node in the network topology. The anchor model adds some inertia to avoid the models at different nodes from drifting away from each other, and reduces the effective ζ of the graph (see Theorem 8.1 for the definition). More details on this variant can be found in [9].

8.3   Error Convergence Analysis

Let us now analyze the effect of the network topology on the error convergence of decentralized SGD. Observe that the consensus step, which mixes the information between neighboring nodes will achieve faster consensus if the underlying graph $\mathcal{G}$ is densely connected. Thus, we expect the error convergence to become worse as the graph becomes more sparse. In this section we provide a convergence analysis that corroborates this intuition.

8.3.1 Assumptions

The convergence analysis requires the following standard assumptions, which we seen in other convergence analyses in this book.

- **Lipschitz Smoothness**: The objective function $F(\mathbf{w})$ is differentiable and L-Lipschitz smooth, that is,

$$\|\nabla F(\mathbf{w}) - \nabla F(\mathbf{w}')\| \leq L\|\mathbf{w} - \mathbf{w}'\| \tag{8.8}$$

- **Unbiased Gradients**: The stochastic gradient $g(\mathbf{w}; \xi)$ is an unbiased estimate of $\nabla F(\mathbf{w})$, that is,

$$\mathbb{E}_\xi[g(\mathbf{w}; \xi)] = \nabla F(\mathbf{w})$$

- **Bounded Variance**: The stochastic gradient $g(\mathbf{w}; \xi)$ has bounded variance, that is,

$$\mathbb{E}_\xi[\|g(\mathbf{w}; \xi)\|^2] \leq \|\nabla F(\mathbf{w})\|^2 + \sigma^2 \tag{8.9}$$

- **Mixing Matrix**: The mixing matrix $\mathbf{M}$ is doubly stochastic. This implies that the largest eigen value of the matrix is 1 and all the other eigen values have magnitude strictly less than 1.

8.3.2 Convergence Analysis of Decentralized SGD

Based on these assumptions, [9] presents the following convergence results for decentralized SGD and its variants decentralized local-update SGD. The proofs of these theorems can be found in the appendix of [9].

Theorem 8.1 (Convergence of Decentralized SGD) *For a L-smooth function, if the learning rate satisfies* $\eta L + \eta^2 L^2(2\zeta^2/(1 - \zeta^2) + 2\zeta/(1 - \zeta)^2) \leq 1$, *and if the starting point is* $\mathbf{w}_1$ *then* $F(\mathbf{w}_t)$ *after t iterations of decentralized SGD is bounded as*

$$\mathbb{E}\left[\frac{1}{t}\sum_{k=1}^{t}\|\nabla F(\mathbf{w}_k)\|^2\right] \leq \frac{2[F(\mathbf{w}_1) - F_{inf}]}{\eta t} + \frac{\eta L\sigma^2}{m} + \eta^2 L^2\sigma^2\left(\frac{1 + \zeta^2}{1 - \zeta^2} - 1\right) \tag{8.10}$$

where $\mathbf{w}_k$ *denotes the averaged model at the k-th iteration and the parameter* $\zeta = \max(|\lambda_2(\mathbf{M})|, |\lambda_m(\mathbf{M})|)$.

The first two terms in (8.10) are identical to the error bound for fully synchronous SGD. The last term is the network error term, which arises because of the imperfect consensus between the nodes due to the decentralized topology and the fact that each node can only communicate with its neighbors. This term increases with the parameter ζ, the second largest (by magnitude) eigen value of the mixing matrix $\mathbf{M}$. Sparser graphs have larger ζ, causing an increase in the error floor. Fully synchronous SGD corresponds to $\zeta = 0$, resulting in the last error term being zero.

Proof The proof of Theorem 8.1 is as follows, and it is based on a more general version presented in the Cooperative SGD paper [9]. Similar to the proof of local-update SGD given in Chap. 6, we define the following terms

- $\mathcal{G}_k = \frac{1}{m} \sum_{i=1}^{m} g(\mathbf{w}_k^{(i)})$, the average of stochastic gradients at the workers at iteration k, and
- $\mathcal{H}_k = \frac{1}{m} \sum_{i=1}^{m} \nabla F(\mathbf{w}_k^{(i)})$, the average of the full gradients at the workers at iteration k.

where $\mathbf{w}_k^{(i)}$ denotes the model at the i-th worker in the network. Using these definitions, and following the same steps as the first part of the proof of Theorem 6.1, we obtain the following error decomposition:

$$\frac{1}{t} \sum_{k=1}^{t} \mathbb{E}[\|\nabla F(\mathbf{w}_k)\|^2] \le \underbrace{\frac{2(F(\mathbf{w}_1) - F_{inf})}{t\eta} + \frac{\eta L \sigma^2}{bm}}_{\text{synchronous SGD error}} +$$

$$\underbrace{\frac{L^2}{mt} \sum_{k=1}^{t} \sum_{i=1}^{m} \mathbb{E}[\|\mathbf{w}_k - \mathbf{w}_k^{(i)}\|^2] - \frac{(1 - \eta L)}{mt} \sum_{k=1}^{t} \sum_{i=1}^{m} \mathbb{E}[\|\nabla F(\mathbf{w}_k^{(i)})\|^2]}_{\text{additional network error}} \qquad (8.11)$$

$$= \underbrace{\frac{2(F(\mathbf{w}_1) - F_{inf})}{t\eta} + \frac{\eta L \sigma^2}{bm}}_{\text{synchronous SGD error}} +$$

$$\underbrace{\frac{L^2}{mt} \sum_{k=1}^{t} \mathbb{E}[\|\mathbf{W}_k(\mathbf{I} - \mathbf{J})\|_F^2] - \frac{(1 - \eta L)}{mt} \sum_{k=1}^{t} \mathbb{E}[\|\nabla F(\mathbf{W}_k)\|_F^2]}_{\text{additional network error}} \qquad (8.12)$$

where $\mathbf{W}_k$ and $\mathbf{G}_k$ are as defined in (8.5) and (8.6) respectively. The matrix $\mathbf{J} = \mathbf{1}\mathbf{1}^\top/(\mathbf{1}^\top\mathbf{1})$ where $\mathbf{1}$ is the all-ones column vector. Thus, every element of the matrix $\mathbf{J}$ is equal to $1/m$. As usual, $\mathbf{I}$ denotes the identity matrix of size $m \times m$, where m is the number of workers. For matrix $\mathbf{A}$, $\|\mathbf{A}\|_F^2$ denotes its Frobenius norm, which is the sum of the squares of each of the elements of the matrix.

Now let us analyze the term $\mathbb{E}[\|\mathbf{W}_k(\mathbf{I} - \mathbf{J})\|_F^2]$ appearing in the additional error. Recursively expanding the term inside the norm, we have:

$$\mathbf{W}_k(\mathbf{I} - \mathbf{J}) = (\mathbf{W}_{k-1} - \eta\mathbf{G}_{k-1})\mathbf{M}(\mathbf{I} - \mathbf{J}) \tag{8.13}$$

$$= \mathbf{W}_{k-1}(\mathbf{I} - \mathbf{J})\mathbf{M} - \eta\mathbf{G}_{k-1}(\mathbf{W}_{k-1} - \mathbf{J}) \tag{8.14}$$

$$= \mathbf{W}_1(\mathbf{I} - \mathbf{J})\mathbf{M}^{k-1} - \eta\sum_{l=1}^{k-1}\mathbf{G}_l(\mathbf{M}^{k-l} - \mathbf{J}) \tag{8.15}$$

$$= -\eta\sum_{l=1}^{k-1}\mathbf{G}_l(\mathbf{M}^{k-l} - \mathbf{J}) \tag{8.16}$$

where (8.14) follows from the doubly stochastic property of the mixing matrix $\mathbf{M}$, which implies that $\mathbf{MJ} = \mathbf{JM} = \mathbf{J}$ and hence $(\mathbf{I} - \mathbf{J})\mathbf{M} = \mathbf{M}(\mathbf{I} - \mathbf{J})$. We get (8.16) because all the workers are initialized by the same model $\mathbf{w}_1$, which implies that the term $\mathbf{W}_1(\mathbf{I} - \mathbf{J}) = 0$.

From this, we can bound the first term in the additional error as:

$$\mathbb{E}[\|\mathbf{W}_k(\mathbf{I} - \mathbf{J})\|_F^2]$$

$$= \eta^2\mathbb{E}[\|\sum_{l=1}^{k-1}\mathbf{G}_l(\mathbf{M}^{k-l} - \mathbf{J})\|_F^2] \tag{8.17}$$

$$= \eta^2\mathbb{E}[\|\sum_{l=1}^{k-1}(\mathbf{G}_l - \nabla F(\mathbf{W}_l))(\mathbf{M}^{k-l} - \mathbf{J}) + \nabla F(\mathbf{W}_l)(\mathbf{M}^{k-l} - \mathbf{J})\|_F^2] \tag{8.18}$$

$$\leq 2\eta^2\mathbb{E}[\|\sum_{l=1}^{k-1}(\mathbf{G}_l - \nabla F(\mathbf{W}_l))(\mathbf{M}^{k-l} - \mathbf{J})\|_F^2] + 2\eta^2\mathbb{E}[\|\sum_{l=1}^{k-1}\nabla F(\mathbf{W}_l)(\mathbf{M}^{k-l} - \mathbf{J})\|_F^2] \tag{8.19}$$

$$\leq 2\eta^2\sum_{l=1}^{k-1}\mathbb{E}[\|(\mathbf{G}_l - \nabla F(\mathbf{W}_l))\|_F^2]\|(\mathbf{M}^{k-l} - \mathbf{J})\|_{op}^2$$

$$+ 2\eta^2\mathbb{E}[\|\sum_{l=1}^{k-1}\nabla F(\mathbf{W}_l)(\mathbf{M}^{k-l} - \mathbf{J})\|_F^2] \tag{8.20}$$

$$\leq 2\eta^2\sum_{l=1}^{k-1}\frac{m\sigma^2}{b}\zeta^{2(k-l)} + 2\eta^2\mathbb{E}[\|\sum_{l=1}^{k-1}\nabla F(\mathbf{W}_l)(\mathbf{M}^{k-l} - \mathbf{J})\|_F^2] \tag{8.21}$$

$$\leq 2\eta^2\frac{m\sigma^2}{b}\frac{\zeta^2}{1 - \zeta^2} + 2\eta^2\mathbb{E}[\|\sum_{l=1}^{k-1}\nabla F(\mathbf{W}_l)(\mathbf{M}^{k-l} - \mathbf{J})\|_F^2] \tag{8.22}$$

where (8.19) follows from the fact that $(a+b)^2 \leq 2a^2 + 2b^2$. Then we get (8.20) by observing that the cross terms are zero, and that for two real matrices $\mathbf{A}$ and $\mathbf{B}$, if $\mathbf{B}$ is symmetric then $\|\mathbf{AB}\|_F \leq \|\mathbf{B}\|_{op}\|\mathbf{A}\|_F$ (see [9] for the proof). To get (8.21) we use a result that the operator norm of $\mathbf{M}^{k-l} - \mathbf{J}$ is bounded above by ζ, where $\zeta = \max(|\lambda_2(\mathbf{M})|, |\lambda_m(\mathbf{M})|)$. Then, we obtain (8.22) by using the bound $\sum_{l=1}^{k-1} \zeta^{2(k-l)} \leq \sum_{l=-\infty}^{k-1} \zeta^{2(k-l)} = \zeta^2/(1-\zeta^2)$.

Next, let us upper bound the second term in (8.22)

$$\mathbb{E}[\|\mathbf{W}_k(\mathbf{I}-\mathbf{J})\|_F^2] - 2\eta^2\frac{m\sigma^2}{b}\frac{\zeta^2}{1-\zeta^2}$$

$$\leq 2\eta^2\mathbb{E}[\|\sum_{l=1}^{k-1}\nabla F(\mathbf{W}_l)(\mathbf{M}^{k-l}-\mathbf{J})\|_F^2] \tag{8.23}$$

$$\leq 2\eta^2\sum_{l=1}^{k-1}\mathbb{E}[\|\nabla F(\mathbf{W}_l)\|_F^2]\|(\mathbf{M}^{k-l}-\mathbf{J})\|_{op}^2+$$

$$\eta^2\sum_{l=1}^{k-1}\sum_{h=1,h\neq l}^{k-1}\|(\mathbf{M}^{k-l}-\mathbf{J})\|_{op}\|(\mathbf{M}^{k-h}-\mathbf{J})\|_{op}(\mathbb{E}[\|\nabla F(\mathbf{W}_l)\|_F^2]+\mathbb{E}[\|\nabla F(\mathbf{W}_h)\|_F^2])$$

$$\tag{8.24}$$

$$\leq 2\eta^2\sum_{l=1}^{k-1}\mathbb{E}[\|\nabla F(\mathbf{W}_l)\|_F^2]\zeta^{2(k-l)}+$$

$$\eta^2\sum_{l=1}^{k-1}\sum_{h=1,h\neq l}^{k-1}\zeta^{2k-l-h}(\mathbb{E}[\|\nabla F(\mathbf{W}_l)\|_F^2]+\mathbb{E}[\|\nabla F(\mathbf{W}_h)\|_F^2]) \tag{8.25}$$

$$\leq 2\eta^2\sum_{l=1}^{k-1}\mathbb{E}[\|\nabla F(\mathbf{W}_l)\|_F^2]\zeta^{2(k-l)} + 2\eta^2\sum_{l=1}^{k-1}\zeta^{k-l}\mathbb{E}[\|\nabla F(\mathbf{W}_l)\|_F^2]\sum_{h=1,h\neq l}^{k-1}\zeta^{k-h} \tag{8.26}$$

$$\leq 2\eta^2\frac{\zeta^2}{1-\zeta^2}\sum_{l=1}^{k-1}\mathbb{E}[\|\nabla F(\mathbf{W}_l)\|_F^2] + \frac{2\eta^2\zeta}{(1-\zeta)}\sum_{l=1}^{k-1}\zeta^{k-l}\mathbb{E}[\|\nabla F(\mathbf{W}_l)\|_F^2]. \tag{8.27}$$

Now we sum over all k from 1 to t to get the following:

$$\frac{L^2}{mt}\sum_{k=1}^{t}\mathbb{E}[\|\mathbf{W}_k(\mathbf{I}-\mathbf{J})\|_F^2]$$

$$\leq \frac{2\eta^2 L^2\sigma^2}{b}\frac{\zeta^2}{1-\zeta^2} + \frac{2\eta^2 L^2}{mt}\sum_{k=1}^{t}\frac{\zeta}{(1-\zeta)}\sum_{l=1}^{k-1}\zeta^{k-l}\mathbb{E}[\|\nabla F(\mathbf{W}_l)\|_F^2] \tag{8.28}$$

$$\leq \frac{2\eta^2 L^2\sigma^2}{b}\frac{\zeta^2}{1-\zeta^2} + \frac{2\eta^2 L^2}{mt}\sum_{k=1}^{t}\frac{\zeta}{(1-\zeta)^2}\mathbb{E}[\|\nabla F(\mathbf{W}_l)\|_F^2] \tag{8.29}$$

Substituting this into (8.12) we have:

$$\frac{1}{t}\sum_{k=1}^{t}\mathbb{E}[\|\nabla F(\mathbf{w}_k)\|^2] \le \frac{2(F(\mathbf{w}_1) - F_{inf})}{t\eta} + \frac{\eta L\sigma^2}{bm} + \frac{2\eta^2 L^2\sigma^2}{b}\frac{\zeta^2}{1-\zeta^2}$$

$$+ \frac{2\eta^2 L^2}{mt}\sum_{k=1}^{t}\frac{\zeta}{(1-\zeta)^2}\mathbb{E}[\|\nabla F(\mathbf{W}_l)\|_F^2] - \frac{(1-\eta L)}{mt}\sum_{k=1}^{t}\mathbb{E}[\|\nabla F(\mathbf{W}_k)\|_F^2] \quad (8.30)$$

If $\eta L + \eta^2 L^2(2\zeta^2/(1-\zeta^2) + 2\zeta/(1-\zeta)^2) \le 1$ then the last two terms can be dropped. And this completes the proof of Theorem 8.1.

8.3.3 Convergence Analysis of Decentralized Local-Update SGD

Theorem 8.1 can be generalized to the decentralized local-update SGD variant, where each node performs τ local updates before communicating with its neighbors, as follows.

Theorem 8.2 (Convergence of Decentralized Local-update SGD) *Starting from* $\mathbf{w}_1$, *for a L-smooth objective function if the learning rate* η *satisfies*

$$\eta L + \frac{\eta^2 L^2\tau^2}{1-\zeta}\left(\frac{2\zeta^2}{1+\zeta} + \frac{2\zeta}{1-\zeta} + \frac{\tau-1}{\tau}\right) \le 1,$$

then $F(\mathbf{w}_t)$ *after t iterations* $(t/\tau$ *rounds) of decentralized local update SGD is bounded as*

$$\mathbb{E}\left[\frac{1}{t}\sum_{k=1}^{t}\|\nabla F(\mathbf{w}_k)\|^2\right] \le \frac{2\left[F(\mathbf{w}_1) - F_{inf}\right]}{\eta t} + \frac{\eta L\sigma^2}{m} + \eta^2 L^2\sigma^2\left(\frac{1+\zeta^2}{1-\zeta^2}\tau - 1\right)$$

$$(8.31)$$

where $\mathbf{w}_k$ *denotes the averaged model at the k-th iteration.*

The proof of Theorem 8.2 is a combination of the proof of local-update SGD given in Chap. 6 and that of decentralized SGD derived above (see [9] for details). When the communication period $\tau = 1$ and $\zeta = 0$, this bound reduces the convergence bound for synchronous SGD. As τ increases, the adverse effect of network sparsity (larger ζ) is magnified because τ multiplies the term $\frac{1+\zeta^2}{1-\zeta^2}$. To illustrate the effect τ and ζ, we plot the last term in (8.31), referred to as the network error bound (NEB) term in Fig. 8.2a. The network becomes sparser as ζ varies along the x-axis. Each line corresponds to a different value of τ. The bottom left corner of the plot corresponds to fully synchronous SGD. Using Fig. 8.2a a system designer can select the (ζ, τ) pair that achieves the target error and also has a smaller runtime per iteration than synchronous SGD. Figure 8.2b shows experimental results for a system of 8 worker nodes used to train the VGG-16 network on the CIFAR-10 dataset. We observe

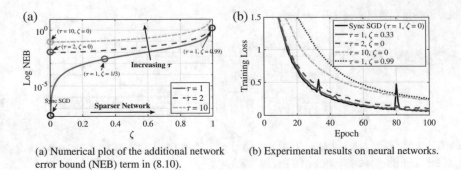

(a) Numerical plot of the additional network error bound (NEB) term in (8.10).

(b) Experimental results on neural networks.

Fig. 8.2 a Illustration of how the additional network error term in (8.31) monotonically increases with τ and ζ; **b** Experiments on CIFAR-10 with VGG-16 and 8 worker nodes. For the same learning rate, larger τ or larger ζ lead to a higher error floor at convergence. Each line in (**b**) corresponds to a circled point in (**a**)

that by using a sparser network, we can achieve almost the same error versus iterations convergence as synchronous SGD.

8.4 Runtime Analysis

In this section, we quantify the runtime advantage of using a sparse network topology to perform consensus between the m worker nodes. The runtime per iteration of decentralized SGD depends on the communication protocol used for inter-node information exchange. Below, we show the analysis for one such communication protocol.

To obtain a simple analysis of how the runtime per iteration scales with the properties of the graph, we assume that the communication between any pair of nodes can happen in parallel. This can be implemented in practice using (1) time division multiplexing, where each edge is assigned a time slot in which nodes connected by that edge can exchange information or (2) frequency division multiplexing, where each edge is assigned a frequency band that they can use to communicate. More efficient communication protocols that reduce the number of time slots or frequencies required can be designed by decomposing the network into matchings. We refer the readers to [10] for a detailed description of such improved communication protocols.

Suppose that local computation time to complete one SGD iteration at each node is Y, assumed to be constant here for simplicity. We also consider that the communication delay in exchanging models between nodes i and j is an exponentially random variable $T_{i,j} \sim \mathrm{Exp}(\mu)$, which is independent and identically distributed across edges. Let us compute the expected time T_i taken by each node i exchange (send and receive) model updates with its neighbors $j \in \mathcal{N}(i)$. The random variable T_i is the maximum of the communication times on each of node i's links:

$$T_i = \max_{j \in \mathcal{N}(i)} T_{i,j} \tag{8.32}$$

$$\mathbb{E}[T_i] \approx \frac{\log d_i}{\mu} \tag{8.33}$$

Since all inter-node communications occur in parallel, the expected runtime per iteration is the maximum of T_i's of all nodes:

$$\mathbb{E}[T] = Y + \mathbb{E}[\max_{i \in \mathcal{V}} T_i] \tag{8.34}$$

$$\geq Y + \max_{i \in \mathcal{V}} \mathbb{E}[T_i] \tag{8.35}$$

$$\approx Y + \frac{\log(\max_i d_i)}{\mu} \tag{8.36}$$

where to get (8.35), we use the convexity of the maximum function to switch the order of the max and the expectation. Thus, the expected runtime is lower bounded by a function that increases with the maximum node degree in the graph. For exponential inter-node communication times, the function scales logarithmically with the max node degree. In contrast, for fully synchronous SGD, recall that the expected runtime per iteration is $\mathbb{E}[T] \approx \frac{\log m}{\mu}$. By using a sparse graph, we can ensure that the max degree $\max_i d_i$ is significantly smaller than m and thus decentralized SGD can achieve a large speed-up over synchronous SGD.

If we perform decentralized local-update SGD with τ local updates at each node, then the communication delay will be amortized across the τ iterations to get

$$\mathbb{E}[T] = Y + \frac{\mathbb{E}[\max_{i \in \mathcal{V}} T_i]}{\tau} \tag{8.37}$$

$$\geq Y + \frac{\max_{i \in \mathcal{V}} \mathbb{E}[T_i]}{\tau} \tag{8.38}$$

$$\approx Y + \frac{\log(\max_i d_i)}{\mu \tau} \tag{8.39}$$

In Table 8.1, we compare decentralized SGD with fully synchronous SGD in terms of the maximum number of handshakes required and the maximum transmitted data size. In fully synchronous SGD, the parameter server needs to communicate with all m nodes and exchange a d-dimensional model $\mathbf{w}$ with each of the m nodes. In decentralized SGD, each node only needs to communicate with d_i neighbors and exchange models with them.

In Fig. 8.3 we show experimental results comparing different decentralized local-update SGD protocols, firstly in terms of the training loss versus epochs (or traversals of the training dataset) and secondly in terms of the training loss versus wallclock time. Performing sparse (higher ζ) or infrequent (large τ) consensus between the nodes results in a worse error versus epochs convergence. However, due to the savings in the wallclock runtime spent per iteration, we can achieve a significant convergence speed-up as seen in the right side subfigure.

Table 8.1 Comparison between sparse averaging (decentralized averaging) and full synchronization (via a parameter server). When the latency to establish handshakes is dominant, sparse averaging can provide significant reduction in communication time

Averaging protocol	# Handshakes	Transmitted data size
Decentralized	$\max_i d_i$	$2d \cdot \max_i d_i$
Fully synchronized (parameter server)	m	$2dm$

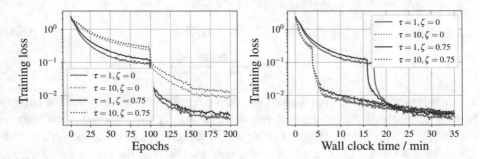

Fig. 8.3 Decentralized periodic averaging on CIFAR-10 with VGG-16. Fully synchronous SGD corresponds to ($\tau = 1, \zeta = 0$). Allowing more local updates (higher τ) leads to slower convergence in terms of epochs. But it requires about 4x less wall-clock time to achieve a training loss of 0.1

Summary

In this chapter we considered a variant of our distributed optimization problem where there is no central parameter server to coordinate between the worker nodes. Instead, the workers are connected via an arbitrary decentralized topology and each worker can only exchange updates with its neighbors. We described the decentralized SGD algorithm and several of its variants that combine decentralized consensus with previously proposed communication-efficiency and variance reduction strategies. We then presented an error convergence analysis and a runtime analysis of decentralized SGD, along with experimental results showing the trade-off between error and runtime.

Problems

1. Consider a network of 4 worker nodes that are connected by a ring topology, and they collaboratively train a machine learning model using decentralized SGD. At every averaging step, assume that each worker assigns weight α to each of its neighbors. Write down the mixing matrix of this system of nodes.

2. Please calculate the ζ value of the mixing matrix that you wrote down in Problem 1 (you can use `numpy.linalg.eig` to find the eigen values). Plot ζ for α taking values from the set $\{0, 0.005, 0.100, 0.105, 0.110, \ldots, 0.500\}$. In order to achieve the lowest error floor at convergence, what is the best value of α? In order to guarantee the convergence, what is range of feasible α?

References

1. J. Tsitsiklis, D. Bertsekas, and M. Athans, "Distributed asynchronous deterministic and stochastic gradient optimization algorithms," *IEEE Transactions on Automatic Control*, vol. 31, no. 9, pp. 803–812, 1986.
2. A. Nedic and A. Ozdaglar, "Distributed subgradient methods for multi-agent optimization," *IEEE Transactions on Automatic Control*, vol. 54, no. 1, pp. 48–61, 2009.
3. K. Yuan, Q. Ling, and W. Yin, "On the convergence of decentralized gradient descent," *SIAM Journal on Optimization*, vol. 26, no. 3, pp. 1835–1854, 2016.
4. X. Lian, C. Zhang, H. Zhang, C.-J. Hsieh, W. Zhang, and J. Liu, "Can decentralized algorithms outperform centralized algorithms? a case study for decentralized parallel stochastic gradient descent," in *Proceedings of the International Conference on Neural Information Processing Systems*, 2017, pp. 5330–5340.
5. K. Scaman, F. Bach, S. Bubeck, L. Massoulié, and Y. T. Lee, "Optimal algorithms for non-smooth distributed optimization in networks," in *Advances in Neural Information Processing Systems*, 2018, pp. 2740–2749.
6. A. Koloskova, S. U. Stich, and M. Jaggi, "Decentralized stochastic optimization and gossip algorithms with compressed communication," arXiv preprint arXiv:1902.00340, 2019.
7. D. Jakovetic, D. Bajovic, A. K. Sahu, and S. Kar, "Convergence rates for distributed stochastic optimization over random networks," in *2018 IEEE Conference on Decision and Control (CDC)*. IEEE, 2018, pp. 4238–4245.
8. M. Assran, N. Loizou, N. Ballas, and M. Rabbat, "Stochastic gradient push for distributed deep learning," in *Proceedings of the 36th International Conference on Machine Learning*, ser. Proceedings of Machine Learning Research, vol. 97. PMLR, 09–15 Jun 2019, pp. 344–353. [Online]. Available: http://proceedings.mlr.press/v97/assran19a.html
9. J. Wang and G. Joshi, "Cooperative sgd: A unified framework for the design and analysis of local-update sgd algorithms," *Journal of Machine Learning Research*, vol. 22, no. 213, pp. 1–50, 2021. [Online]. Available: http://jmlr.org/papers/v22/20-147.html
10. J. Wang, A. Sahu, G. Joshi, and S. Kar, "MATCHA: Speeding Up Decentralized SGD via Matching Decomposition Sampling," *preprint*, May 2019. [Online]. Available: https://arxiv.org/abs/1905.09435
11. B. Li, S. Cen, Y. Chen, and Y. Chi, "Communication-efficient distributed optimization in networks with gradient tracking and variance reduction," in *Proceedings of the International Conference on Artificial Intelligence and Statistics (AISTATS)*, ser. Proceedings of Machine Learning Research, vol. 108. PMLR, 26–28 Aug 2020, pp. 1662–1672. [Online]. Available: https://proceedings.mlr.press/v108/li20f.html

Beyond Distributed Training in the Cloud 9

Let us summarize the concepts that we learned through this book, and discuss the future of distributed machine learning beyond cloud-based implementations that we studied in this book.

We started off this book by analyzing the convergence of classic single-node SGD and studied some of its variance-reduced variants such as SAG and SAGA. To handle large training datasets in a fast and efficient manner, practical implementations require SGD to be run in a distributed manner. The main focus of this book was on studying distributed stochastic gradient descent (SGD) algorithms, which are used to perform machine learning training using worker nodes that are servers in the cloud. In Chap. 4, we introduced the first and the most common distributed SGD algorithm, synchronous SGD, which uses a parameter server and a system of worker nodes, which are servers in the cloud, and we analyzed its convergence and runtime per iteration. To overcome tail latency due to straggling workers in synchronous SGD, in Chap. 5 we introduced the asynchronous SGD algorithm, which removes the synchronization barrier and allows worker nodes to independently communicate with the parameter server whenever they finish their local computation. Then in Chaps. 6 and 7 we studied two ways to reduce the communication cost of sending and receiving model updates between the worker nodes and the parameter server. Finally in Chap. 8 we discussed decentralized SGD using an arbitrary peer-to-peer topology connecting worker nodes, without having a central parameter server to aggregate their updates. A key insight from all these algorithms and their analyses was that the standard optimization-theoretic approach that seeks to speed-up the error versus iterations convergence is insufficient to capture the true convergence speed with respect to wallclock time. The design of distributed SGD algorithms has to take into account how the synchronization and communication protocol affects the wallclock runtime spent per iteration. This book advocates such a system-aware philosophy in the design of distributed machine learning.

In most modern machine learning applications, edge nodes such as cellphone or sensors collect rich training data from their environment which can be used for data-driven decision-

G. Joshi, *Optimization Algorithms for Distributed Machine Learning*,
Synthesis Lectures on Learning, Networks, and Algorithms,
https://doi.org/10.1007/978-3-031-19067-4_9

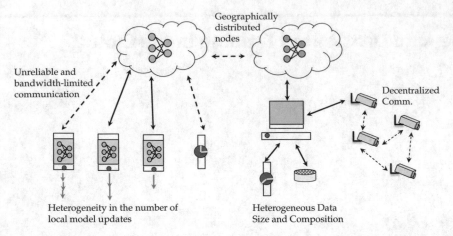

Fig. 9.1 Federated learning using communication-limited edge devices such as cellphones, smart-watches, IoT cameras, etc. Computational speeds and the size and composition of data can vary widely across nodes

making. If we wish to employ cloud-based training using the parameter server framework then these raw training data will have to transferred to the cloud, where it can shuffled and partitioned across worker nodes. However, due to limited communication capabilities as well as privacy concerns, it is prohibitively expensive to send the data collected by edge nodes to the cloud for centralized processing. The nascent research field called federated learning [1, 2] considers a large number of resource-constrained edge clients such as cellphones or IoT sensors that collect training data from their environment. Instead of sending the training data to the cloud, the nodes locally perform a few iterations of training and only send the resulting model to the cloud, as illustrated in Fig. 9.1. While the federated learning algorithms are similar to local-update SGD and other communication-efficient distributed SGD algorithms, where are some unique aspects of the large-scale federated setting that requires innovations to cloud-based distributed training algorithms.

1. **Data Heterogeneity**. Unlike cloud-based training where a central training dataset is shuffled and then evenly partitioned across the worker nodes, in federated learning, the data is collected independently by each edge client. Thus the local datasets are highly heterogeneous across the clients both in terms of size and distribution. This data heterogeneity when combined with limited communication between the edge clients and the cloud can result in a higher error floor. Recent works [3, 4] have proposed techniques to combat such data heterogeneity.
2. **Computational Heterogeneity**. In the cloud-based training setting, we considered homogeneous worker nodes that can experience short-term delay fluctuations. However, in the federated learning setting, the computational capabilities can vary widely across

edge clients due to different device types (phones versus sensors) and varying hardware speeds. The amount of local computation can also vary due to the heterogeneity in the local datasizes. For example, in federated learning, each node performs E $epochs$ (traversals of their local dataset). If a client has n_i local data samples and the mini-batch size is b, the number of local SGD iterations is $\tau_i = En_i/b$, which can widely vary across nodes. Alternately, if we fix a time interval instead of fixing the number local epochs, then faster nodes can perform more local updates than slower nodes within the same time interval. Finally, the client may also use different local optimizers, which can result in heterogeneous local progress. Some recent works [5–7] propose new algorithms to tackle such data heterogeneity.

3. **Communication Constraints**. Since edge clients are wirelessly connected to the cloud and to the other clients, typically with bandwidth-limited links, the distributed training algorithms have operate under much stricter communication constraints than cloud-based training. Techniques such local updates, quantization and sparsification of updates are essential to enable federated training algorithms to scale to a large number of devices.

4. **Intermittent Availability**. Due to the sheer scale of federated learning where thousands or even millions of geographically distributed edge clients can be involved in the training of a machine learning model, it is infeasible for all clients to be available in each training round. The availability of devices depends on their timezone, their battery status and network connectivity. Therefore, federated learning algorithms have to allow only a subset of clients to participate in each training round. Recent works such as [8] propose client selection strategies and others such as [9] try to reduce the variance due to partial client participation.

5. **Private and Secure Aggregation**. Although the data stays on the edge client, the model updates that are communicated with the cloud can reveal private information about the users' data. Local and global differential privacy techniques [10] can be used to obfuscate local updates sent by the clients. However, there is a trade-off between the convergence speed and the privacy offered by these methods An alternative to privacy is to use cryptographic secure aggregation protocols [11]. However, such secure aggregation protocols require additional computation, which make them hard to scale to a large number of edge clients.

6. **Adversarial Nodes**. In the data-center setting, all the worker nodes are under centralized control by the system administrator and it is difficult for malicious adversaries to take control of one or more worker nodes. However, with thousands of edge clients in the federated learning scenario, adversaries can control a small subset of devices without being identified by the central aggregating server. Some recent works such as [12, 13] show that even a single adversarial client can send poisonous updates and replace the trained global model by any model of its choice. Currently, federated learning algorithms employ clipping of clients' updates in order to mitigate the effect of adversarial clients [14].

Although we did not cover federated learning and the above aspects in this book, the foundation of system-aware distributed algorithm design established can be helpful in the design of federated optimization algorithms.

References

1. H. B. McMahan, E. Moore, D. Ramage, S. Hampson, and B. A. y Arcas, "Communication-Efficient Learning of Deep Networks from Decentralized Data," *International Conference on Artificial Intelligenece and Statistics (AISTATS)*, Apr. 2017. [Online]. Available: https://arxiv.org/abs/1602.05629
2. P. Kairouz, H. B. McMahan, B. Avent, A. Bellet, M. Bennis, A. N. Bhagoji, K. Bonawitz, Z. Charles, G. Cormode, R. Cummings, R. G. L. D'Oliveira, S. E. Rouayheb, D. Evans, J. Gardner, Z. Garrett, A. Gascon, B. Ghazi, P. B. Gibbons, M. Gruteser, Z. Harchaoui, C. He, L. He, Z. Huo, B. Hutchinson, J. Hsu, M. Jaggi, T. Javidi, G. Joshi, M. Khodak, J. Konecny, A. Korolova, F. Koushanfar, S. Koyejo, T. Lepoint, Y. Liu, P. Mittal, M. Mohri, R. Nock, A. Ozgur, R. Pagh, M. Raykova, H. Qi, D. Ramage, R. Raskar, D. Song, W. Song, S. U. Stich, Z. Sun, A. T. Suresh, F. Tramer, P. Vepakomma, J. Wang, L. Xiong, Z. Xu, Q. Yang, F. X. Yu, H. Yu, and S. Zhao, "Advances and open problems in federated learning," *Foundations and Trends in Machine Learning*, Jun. 2021. [Online]. Available: https://arxiv.org/abs/1912.04977
3. X. Li, K. Huang, W. Yang, S. Wang, and Z. Zhang, "On the convergence of fedavg on non-iid data," in *International Conference on Learning Representations (ICLR)*, Jul. 2020. [Online]. Available: https://arxiv.org/abs/1907.02189
4. S. P. Karimireddy, S. Kale, M. Mohri, S. J. Reddi, S. U. Stich, and A. T. Suresh, "SCAFFOLD: Stochastic controlled averaging for on-device federated learning," *arXiv preprint* arXiv:1910.06378, 2019.
5. J. Wang, Q. Liu, H. Liang, G. Joshi, and H. V. Poor, "Tackling the Objective Inconsistency Problem in Heterogeneous Federated Optimization," in *Proceedings on Neural Information Processing Systems (NeurIPS)*, Dec. 2020. [Online]. Available: https://arxiv.org/abs/2007.07481
6. J. Wang, Z. Xu, Z. Garrett, Z. Charles, L. Liu, and G. Joshi, "Local adaptivity in federated learning: Convergence and consistency," *arXiv preprint* arXiv:2106.02305, 2021.
7. F. Mansoori and E. Wei, "Flexpd: A flexible framework of first-order primal-dual algorithms for distributed optimization," *IEEE Transactions on Signal Processing*, vol. 69, pp. 3500–3512, 2021.
8. Y. J. Cho, J. Wang, and G. Joshi, "Client selection in federated learning: Convergence analysis and power-of-choice selection strategies," 2020.
9. X. Gu, K. Huang, J. Zhang, and L. Huang, "Fast federated learning in the presence of arbitrary device unavailability," *Advances in Neural Information Processing Systems*, vol. 34, 2021.
10. K. Wei, J. Li, M. Ding, C. Ma, H. H. Yang, F. Farokhi, S. Jin, T. Q. S. Quek, and H. Vincent Poor, "Federated learning with differential privacy: Algorithms and performance analysis," *IEEE Transactions on Information Forensics and Security*, vol. 15, pp. 3454–3469, 2020.
11. K. Bonawitz, V. Ivanov, B. Kreuter, A. Marcedone, H. B. McMahan, S. Patel, D. Ramage, A. Segal, and K. Seth, "Practical secure aggregation for federated learning on user-held data," in *NIPS Workshop on Private Multi-Party Machine Learning*, 2016.
12. H. Wang, K. Sreenivasan, S. Rajput, H. Vishwakarma, S. Agarwal, J.-y. Sohn, K. Lee, and D. Papailiopoulos, "Attack of the tails: Yes, you really can backdoor federated learning," *Advances in Neural Information Processing Systems*, vol. 33, pp. 16 070–16 084, 2020.

13. E. Bagdasaryan, A. Veit, Y. Hua, D. Estrin, and V. Shmatikov, "How to backdoor federated learning," in *International Conference on Artificial Intelligence and Statistics*. PMLR, 2020, pp. 2938–2948.
14. Z. Sun, P. Kairouz, A. T. Suresh, and H. B. McMahan, "Can you really backdoor federated learning?" *arXiv preprint* arXiv:1911.07963, 2019.

Printed in the United States
by Baker & Taylor Publisher Services